The Greatest Mysteries of all Times

Deceptions Between Science and Science Fiction

Pierre A. Kandorfer, Ph.D.

Dedicated to my wife Helga, my daughter Michelle Dominique, my granddaughter Diana Maria, my grandson Luca Norman, and my great granddaughter Lea Maria

Chapters

Helium and hydrogen are too complex
How can helium and hydrogen clump together?

Why is earth's center still
molten hot?
The zircon-lead and helium
rations
The soil-water ratio
The Niagara falls
The river path
The ocean floor
The Sequoias tell their story
What historical records reveal
The solar eclipses
The written records
The languages spoken
Are star clusters moving away?
Not enough hydrogen
Why is hydrogen still left?
The shrinking sun problem
Sun emitting solar neutrinos
Comets circling the sun
Where is the water?
The solar wind radiation
The Saturn ring phenomenon
The Galilean moons
The radioactive decay
The carbon 14 disintegration
factor
Where are the meteorites?
Meteor craters at earth's
surface

How did DNA originate?
How could the mutation create a new species?
How did the multi-cellular life originate?
How did sex originate?
Where are the missing transitional fossils?
How can living creatures remain unchanged over millions of years?
How can blind chemistry generate factors such as mind, intelligences, meaning, or morality?
Where are the scientific breakthroughs of the evolution?
Why is creationism banned from our schools?

The greatest mysteries of all times

Our secular society wants us to believe that everything in our universe is material. Our entire existence is supposed to be based virtually only on what we can see, touch, or physically manipulate. This atheistic worldview claims that the fundamental building block of everything is just matter.

In short, our entire life is only what natural sciences are able to rationally explain. There is no higher power, individualism, spirituality, or supernatural phenomenon. Nothing that goes beyond pure physical existence. Creationism, as documented in the Bible, is just a legend.

Evolutionism is the general idea that our universe is the result of "random cosmic accidents." It says, life arose spontaneously by chance through accidental chemical processes, and all life-forms are related and share a common ancestor. All living creatures have one forefather: from bananas to birds, from fishes to flowers, from apes to humans, etc.

As the enlightenment period, rationalism, and evolution theory suggested, most traditional social, religious, and political ideas must be rejected. This worldview is perfectly in sync with the evolution theory. The movement was described as "scientific" as opposed to creationism, which is still often labeled as "unprovable" and a religious dogma.

The sciences and rationalism were supposed to explain everything, but can they? Despite countless outrageous, shameful, and dishonest claims, evolution, materialism, and rationalism leave us with more questions than answers.

The supposedly "scientifically provable" evolution, the big bang, and scientific rationalism theories remain twisted. These deceitful theories are defying common sense, logic, and any reliable, methodical proof every honest scientific undertaking relies upon.

What is left? Thousands of theories with no proof, confirmation, or logical explanation. They are nothing but the greatest mysteries of all time.

For more than a century, most scientific discoveries and explanations have been made by a mid-19th century theory described as evolution, which contributed to a series of other theories with no foundation in scientific facts at all.

The creationism, as taught by the Bible, is outlawed in our public schools, but the evolution theory is not. Just the opposite, it is the official description of where and how we came from in our entire educational system. As inappropriate it may sound, the evolution theory is considered based on "scientific facts." Really?

Vance Farrell describes in his excellent paperback "The evolution handbook" what evolutionists believe in:

- Our system of the universe with galaxies, stars, planets, and moons was created out of nothing plus time, billions of years.
- Living creatures were supposedly generated out of dirt, water, and lots of time.

Considering our natural laws, this is laughable at best and incredibly dishonest at worst. Thank God, it is mostly easy to differentiate between science and science fiction.

Mystery #1
The origin of everything

Secular physicists argue that our universe created itself "out of noting." They call this phenomenon the "Heisenberg uncertainty principle."

This evolutionary formula for the creation of everything out of nothing is simple. According to this theory, nothing plus nothing produces two elements. Adding time creates further 92 elements known to us. By adding more time, all physical laws and the completely structured, astoundingly complex system of the universe with galaxies, stars, planets, and orbiting moons spring into their existence in perfect order. This all, of course, happened just by accident, as a result of accidental mishaps, random confusion, and time. Billions of years.

Over one hundred years, world-famous people demand to forget the Bible and the history of Creation because it is "harmful to the society."

"Religion is an attempt to get control over the sensory world, in which we are placed, using the wish-world, which we have developed inside us as a result of biological and psychological necessities. But it cannot achieve its end. Its doctrines carry with them the stamp of the times in which they originated, the ignorant childhood days of the human race. Its consolations deserve no trust. Experience teaches us that the world is not a nursery. The ethical commands, to which religion seeks to lend its weight, require some other foundations instead, for human society cannot do without them, and it is dangerous to link up obedience to them with religious belief. If one attempts to assign to religion its place in man's evolution, it seems not so much to be a lasting acquisition, as a parallel to the neurosis which the civilized individual must pass through on his way from childhood to maturity." (Sigmund Freud)

Is this explanation logical? Does it make any sense? Is this theory scientifically verifiable?

Among atheists, this doesn't really matter. The only thing that matters for modern-day progressives is finding a theory that makes the divine creation obsolete. Insisting there is no God is for them more important than scientific truth.

"Evolution is promoted by its practitioners as more than mere science. Evolution is promulgated as an ideology, a secular religion, a full-fledged alternative to Christianity, with meaning and morality… Evolution is a religion. This was true of evolution in the beginning, and it is true of evolution still today." (Michael Ruse, professor of philosophy and zoology)

Mystery #2
The origin of life

The origin of life is one of the greatest mysteries of all time. According to most school books, life was created from dirt, water, and a lot of time through the so-called "primordial soup" around four billion years ago.

This "chemical evolution" is one of the most disturbing theories in the entire scientific world. It offers no answers, just questions.

According to the secular evolution theory, life on earth was created about five billion years ago and gradually progressed through a series of stages.

Stage 1
Evolutionists suggest that our atmosphere was very different from today. Our atmosphere contains 21 percent oxygen, 78 percent nitrogen, and one percent of other gases. The anti-genesis followers, however, hypothesize that at that time our atmosphere was made up of methane, ammonia, hydrogen, and water vapor.

Stage 2
It is estimated that the ultraviolet light, electric discharge, and a "high-energy bombardment" lead to the creation of small organic molecules such as amino acids and nucleotides.

Stage 3
It is imagined that small molecules started producing large polymers like proteins and nucleic acids such as DNA. Nobody knows how.

Stage 4

It is theorized that now these large molecules clumped together and created a gel-like substance they call microspheres or coacervates. They supposedly attracted smaller molecules and produced new organic structures biology calls protocells.

Stage 5

Suddenly, some of the complex molecules established somehow the ability to "duplicate" themselves and became "living cells." As soon as the oxygen in the atmosphere increased, "most forms of life we know today started to appear."

Extraterrestrial life

Not surprisingly, none of these theories can be proven or replicated in a laboratory setting. This energized many scientists sharing the atheist worldview to focus on a potential extraterrestrial life that would somehow limit their embarrassment with the failing evolution theory.

During the last decade, billions have been spent to study "bio-astronomy" in the hope to find extraterrestrial life. Over a dozen "exobiology" projects studying life in outer space have been conducted, totaling billions of dollars, just in the US.

The Caltech physicist Hugh Ross contemplates that solar winds may be able to transport particles of previously living substances through space. Ross concludes some of them may have landed on Mars. So far, nothing has ever been detected.

Never recreated
Despite thousands of laboratory attempts, the process of chemical evolution was never recreated.

Incredibly complex
Even the simplest organisms of nature are so incredibly complex that they are far beyond any random mathematical chance we can imagine.

No life from death chemicals
The nature is incapable to randomly create the simplest living organic molecules out of dead chemicals. Even the very simplest bacteria are highly complex in nature.

Irreducible complexity
Such a process depends on the "irreducible complexity" and requires highly sophisticated chemical functions. This means that everything must be in perfect order to succeed. Every cell must be absolutely complete to duplicate itself and survive.

The perfect example of the irreducible complexity is the bacterial flagellum, which is technically working similar to a rotating motor with dozens of separate but perfectly synchronized parts. Only a highly intelligent, supreme designer can create such a complex machine.

An over-simplified analogy example might be a sophisticated rocket with thousands of parts and technical components all working together to make the missile fly. You can only make it work if all pieces needed are at the right place at the right time and all procedures work precisely according to the requirement. If one single screw is missing or is at the wrong place at the wrong time, nothing is working.

Incompatible with the law of thermodynamics
The chemical evolution defies the law of thermodynamics, which makes the process impossible.

Physically and mathematically unworkable
The chemical evolution is physically and mathematically impossible. The proposed time factor, supposedly billions of years, doesn't change anything.

Mystery #3:
The Big Bang theory

The most crucial question of all times is how and when our universe was created. How did it reach the infinite place we observe today? Secularists claim that the consensus among physicists, astronomers, and cosmologists is that "the universe as we know was created in a massive explosion that not only created the majority of matter but the physical laws that govern our ever-expanding cosmos."

This is known as The Big Bang theory. With no basis in logic and scientific verification, this is the leading model taught in virtually all schools and colleges worldwide.

For people who want to deny the knowledge of biblical creation, the Big Bang theory sounds quite simple. The Big Bang hypothesis states that all of the current and past matter in the universe came into existence at the same time. Their estimate is around 13.8 billion years ago.

At that time, "all matter was compacted into a very small ball with infinite density and intense heat called singularity." The phenomenon of singularity "began to expand and the universe has suddenly created the way we know it today." Boom! That's it!

Evolution supposedly happened in different phases:

- Cosmic evolution through the Big Bang.
- Stellar evolution is also called chemical evolution or planetary evolution. This should explain how hydrogen and helium, produced

by the explosion at the Big Bang, supposedly created stars and planets.

- Organic evolution claims to explain how life was spontaneously created through "prebiotic soup" (dirt and water).
- Macroevolution claiming that all living organisms have the same ancestor.
- Microevolution intends to explain variations among living creatures.

If you take the natural laws of physics into account, the evolution theory sounds like a fairy tale in a Disney movie. Curious, educated people deserve an explanation.

How can "nothing" produce something?
How can a bit of nothing, packed tightly together, blow up to produce all the matter of the entire universe? Also, how can a "nothingness" pack together?

The vacuum has no density
According to the laws of physics, a vacuum has no density. How can it become so dense to explode? The vacuum is the opposite of the total density. How can science explain this?

No explosion without ignition
How can nothingness explode without any kind of ignition? How can be this a chemical explosion if there are no chemicals present? It also can't be any nuclear explosion without any existing atoms.

"Nothing" can't expand
How can "everything expand" if there is nothing there? Some kind of gravity, assuming it was present,

would keep everything tightly together instead of expanding.

Nothing can't generate heat
How can nothingness generate heat? Where does the heat come from?

Mathematical calculations beyond our skills
How can such a process, assuming it would be physically possible, produce mathematical calculations beyond our imagination and capabilities?

Where does the antimatter come from?
Where would the antimatter come from necessary for such a hypothesis to be feasible?

Antimatter would destroy actual matter
Wouldn't the antimatter, generated by a Big, Bang destroy all the actual matter?

Outward push can't unite particles
How can the process unite the particles being pushed outward?

How can particles slow down?
How can the proposed particles ever slow down?

How can particles change speed or direction?
How can they ever change if their speed and direction remain constant forever?

Helium and hydrogen are too complex
How can atomic structures originate if even just hydrogen and helium are very complex in nature?

How can hydrogen and helium "clump" together?

How can hydrogen and helium clump together in space?

Mystery #4
The creation of stars

Stars fascinated humans from the very beginning of our history. We want to know everything about them, especially when and how they were created. All cultures have their own mythological narrative about stars. The history is full of mesmerizing tales about stars, their alleged meaning, and assumed symbolism.

According to secular scientists, stars were formed from large clouds of gas. They might have been remainders of exploding older stars, or leftovers from the Big Bang, very cold but not very dense.

"Gravity, which causes all the atoms of the gas to pull each other together, makes the cloud contract. As the cloud gets smaller, the atoms of the gas get closer together and begin to bump into each other more, which heats them up. The cloud keeps contracting and getting hotter until the pressure from the heat (which pushes the atoms apart) balances the force of gravity (which pulls the atoms together). When this happens, the cloud is a stable ball of gas like our sun, and is hot enough that it glows - it is a star," the UCSB Science Line argues.

There are several variations of the gas hypothesis described by the University of California Santa Barbara Science Line.

- Stars are formed from space dust and gases and called "interstellar medium." It is so cold in space that "the dust slowly comes together and after millions or billions of years the dust is so dense that the temperature causes it to undergo nuclear reaction and a star is formed."

- "A star is formed when a cloud of gas and dust in space has enough mass that the gravity from that mass pulls the material of that cloud to a central point rather than the particles of gas and dust escaping by their own velocity. As this material collects to a central point, atoms and molecules start bumping into each other. As more and more material assembles, the material bumps together with higher and higher energy, making heat. With the still further collection of material to the center of mass, the energy of hydrogen atoms bumping into one another becomes so great that they will occasionally fuse together to form a helium atom." Such a process is called nuclear fusion and can only take place "where there is a lot of energy to bring these atoms together. "

- "Stars are formed from clouds of gas that are big and dense enough to collapse due to the gravitational forces in the gas cloud. As the gas cloud collapses, the gas heats up and starts to radiate. The gravitational energy is turned into heat and then light. The gas cloud continues to collapse and the temperature increases until the pressure on the inside is high enough to counteract gravity."

Assuming these explanations are the very best science has to offer, are they also at the same time logically, technically, and scientifically feasible? Rationally thinking astrophysicists refer to the natural laws of physics and chemistry.

Gas in space does not clump
Gas in outer space does not clump and can't build up enough mutual gravity to pull itself into stars. Gas floating freely in a vacuum is not able to stick together to form a ball. Accordingly, no gravity can be produced.

Not enough gas matter
Considering this theory, there is by far not enough gas matter available to generate a star.

Not enough time
At the current expansion rate of the universe, there is not enough time for the entire process.

Gas clouds can't contract
Gas clouds in outer space expand, they can't contract.

Gas clouds can only move in one direction
Under these conditions, gas clouds move only in one direction and are too stable to do anything but continuing their one-directional movement.

Not enough mass
There is not enough mass to make these theories physically possible.

No force of attraction
Hydrogen gas in space cannot clump together. There is no force in interstellar space to attract particles to stick to each other

Mystery #5
Exploding stars

According to the Big Bang theory, exploding stars generate heavy metals. "Heavy elements form through a chemical process known as the rapid neutron capture process, or the 'r-process.' But sometimes the heaviest elements, such as plutonium, do not form during this process; it seems that matters get 'stuck' at making elements that are not as heavy, like gold. By seeing one explosion, we can't really resolve the mystery of how the heaviest elements form. That's why we need to see more of them," Ashley Villar, an astrophysicist with a Ph.D. from Harvard University writes.

"Elements that are heavier than hydrogen, such as iron or gold, live in the bellies of stars. And the only way those elements enter and enrich our universe is when stars explode," Villar continues. She studies exploding stars, which are known as supernovae.

No heavy elements from hydrogen and helium
Assuming the Big Bang as a scientifically settled fact, it still only produces hydrogen and helium. How can they generate the ninety other heavier elements?

Change is scientifically impossible
How can hydrogen and helium change itself into one of the heavier elements when their gaps at an atomic weight of 5 and 8 make it scientifically impossible?

Not enough supernova explosions
A normal star is a gigantic ball of gas, about a million times more massive than the earth. Stars are potentially stable for a long time because the energy

produced by the core creates enormous outward pressure. This power balances the inward force of gravity on its huge mass.

Still, when nuclear fuel runs out, there is no longer any force to balance its gravity. Very massive stars mostly collapse very fast, in just about two seconds. This explosion releases a huge amount of energy. Every supernova will out-shine all the billions of other stars in its galaxy.

The supernova collapse is so violent that the electrons and nuclei are crushed together and generate a core of neutrons. This core is so incredibly dense that a teaspoonful would weigh 50 thousand million tons on earth.

The matter cannot be compressed any further, astronomers say. This means that the incoming material from the rest of the star meets a "solid wall " and bounces off the core, rushes outward, and shines extremely brightly. The remaining part of the core is just about 13 miles in diameter and is named a neutron star. It is spinning very fast and has a strong magnetic field. We observe it with regular radio pulses and the object is called a pulsar.

However, how can this theory work if there were not enough supernova explosions until today? Such explosions are extremely rare. Even the most distant stars, supposedly created at the Big Bang billions of years ago, never exploded but still contain heavy elements. Why?

According to physics, there were not enough supernova explosions to build ample new stars as required by the theory. How can the explosions create heavy metals if only hydrogen and helium are observed?

Exploding stars don't produce new stars

How can exploding stars produce new stars if this process does not seem to be technically and scientifically possible?

Mystery #6:
Stellar evolution

According to the evolution theory, the first chemical elements heavier than hydrogen, helium, and lithium formed in nuclear reactions at the centers of the first stars. Much later, when these stars exhausted their fuel of hydrogen and helium, they exploded as supernovas, throwing out the heavier elements.

"These elements, after being transformed in more generations of stars, eventually formed asteroids, moons, and planets. But, how did those first stars of hydrogen and helium form? Star formation is perhaps the weakest link in stellar evolution theory and modern big bang cosmology. " (Rod Bernitt)

Evolutionists believe the star formation began soon after the Big Bang. The cosmic microwave background supposedly started after approximately 300,000 years, the first stars one to ten million years later. Why?

Stellar evolution is one of the most implausible theories in astrophysics. "Stellar evolution is the process in which the forces of pressure (gravity) alter the star. With these forces acting upon stars, their characteristics change dramatically over the period of their existence," the University of Michigan website proclaims.

They argue stellar evolution is "inevitable as stars deplete their initial fuel sources. The search for new fuel sources affects the properties of stars as they evolve. This evolution is a process that consists of many different stages with fuel consumption as the dominant life cycles of an evolving star."

Arthur Holland and Mark Williams conclude, "Stellar evolution, in the form of these fuel consumption stages and their finality, is important because it is responsible for the production of most of the elements (all elements after hydrogen and helium). Moreover, stages in the life cycle of stars are a vital part in the formation of galaxies, new stars, and planetary systems."

This theory leaves us with more questions than answers.

Same amounts of heavy elements
How can be explained that the older stars have no more heavy elements than the youngest ones?

Many stars are still inside of the perimeter
The evolution theory claims the continuing blast of particles outward throughout the space, but there are many stars still in the inside of the outflowing perimeter. How can that happen when it is physically impossible?

Angular momentum
Exploding stars distribute the matter in a strictly linear motion, but many of them have changed their kinesis into spinning, revolving, or orbiting movement (angular momentum). How can this make any scientific sense?

- How can an inward-pushing gas change to a rotating star?
- Why do some stars spin?

- Why do some stars orbit backward?
- Why are some stars traveling much too fast to make the stellar evolution theory feasible?

More questions than answers
Why does the universe show countless stars and just a little gas, while it should be full of gas and only some stars? Why is the universe saturated with superclusters? If water is crucial for the formation of stars, how were the first stars created when no water was available?

Solar energy is generated by the solar collapse
The solar energy appears to be generated by the solar collapse, not nuclear fusion. Wouldn't this make the entire Big Bang theory obsolete?

Mystery #7:
Origin of our solar system

Progressive, evolutionary astronomers claim that the solar system was formed by natural processes about 4.5 billion years ago. For decades, they have been trying to model that formation process using powerful computer simulations. The research has shown that the inner rocky planets and the asteroid belt of our solar system cannot originate naturally at the same time.

"Standard planet-formation models have been unable to reconstruct the distributions of the solar system's small, rocky planets and asteroids in the same simulation." (Nature)

According to astrophysical research, the origin of our solar system has a number of competing secular theories. The "History of cosmological theories" describes them in detail.

The nebular hypothesis
The nebular hypothesis claims "swirling gas" created the sun and the planets and the moons. However, the composition of the planets is very different from each other. How can physics explain that? Only a fast rotating sun might be able to toss off the planets, but the sun rotates very slowly.

The "fission" theory
The "fission" theory wants us to believe that the sunburst sent out the planets and the moons. In this case, the planets would not change their outward trajectory and start circling the sun as they do.

The capture theory

The capture theory hypothesizes that the sun accidentally "captured" the wandering planets and moons. We never observed such an undertaking in space. If they were really pulled by the gravity of the sun, they would crash into it and not circle around.

The accretion theory

The accretion theory suggests that "small chunks of matter gradually pulled together and formed a planet." No such chunks have ever been detected.

The planetary collision theory

The planetary collision theory means that the earth collided with a small planet and a part broke away and produced the moon. How could the moon suddenly start circling the earth? Is this theory applicable to the other 150 moons in space, too?

Colliding planets starting to circle the sun?

The stellar collision theory says that two stars collided and created other planets and moons. How could they start circling the sun?

How can gas planets start circling the sun?

The gas cloud theory emphasizes "gas clouds were pulled from outer space by sun's gravity and produced the planets in our solar system." There is also no explanation though how they suddenly started circling the sun.

Mystery #8:
The balance of the universe

For most people, our universe appears extremely stable, settled, and balanced. In reality, it is just the opposite. The system of the universe is unbelievably fragile, meticulous, and delicate. It depends on thousands of factors, precisely calibrated to make it work. Nothing, absolutely nothing, appears to be coincidental. Everything seems to be designed to the tiniest fraction humanly imaginable.

The nuclear forces
If the nuclear forces were just a portion of a hundred stronger or weaker, carbon could not exist - and so neither couldn't we. A slight increase of only two percent would also automatically eliminate protons.

The gravity
The gravity in space is extremely carefully calibrated. Only a slight change would prohibit stars to form and automatically eliminate the existence of heavy elements.

The most critical balance
The proton and neutron ratio is exactly predetermined. The mass of the neutron must exceed that of the proton to enable the elements to exist. The critical balance is only one part in a thousand. If the ratio of the subatomic particles would exceed that tiny limit, everything would be destroyed.

The proton-baryon ratio
The photon to baryon ratio also is extremely finely tuned. Baryon is a subatomic particle with an equal or

greater weight than a proton. If the ratio is higher than required, stars and galaxies could never hold together through their gravitational attraction.

The electromagnetic force
The electromagnetic force too is essential for the functioning of the universe. If this force would be greater or smaller by a factor of 1.6, all existing protons would decay into leptons.

Three times larger electromagnetic charge of the electron would make any element impossible to exist, with the exception of hydrogen. This would also result in the destruction of all neutral atoms at the temperature of the outer space.

The nebular hypothesis
The widely accepted evolutionary view of the origin of the solar system is usually called the Nebular Hypothesis. This idea suggests, a giant cloud in space made up of mainly spinning, ionized gas with a magnetic field is believed to have pulled together by gravity into the sun, planets and other objects in our solar system.

The computer simulations of this process do not start with initial conditions like those of real nebulas, and have other problems. One scientist summarized these by saying "The clouds are too hot, too magnetic, and they rotate too rapidly." Some scientists described this hypothesis as "the best fit" considering the evidence we can observe.

Dr. Jonathan Henry (Clearwater Christian College) considers solar system evidence to be consistent with a universal catastrophe, which he associates with God's curse on all creation.

Computer simulations favor evolution

The starting condition of most computer simulations purposely favors evolution. They even don't start with the gas and dust-filled nebular cloud, which supposedly is the very beginning of the solar system. It is assumed, planet-size bodies are already formed from the "accumulation of mass." Nobody knows how.

The computer simulation starts with ten or twenty "planets" and thousands of small solid objects of debris (planetesimals) up to a few hundred miles across. Now, the computer simulations are "allowed to evolve" under standard gravitational physic with various initial parameters in an effort to produce the solar system we observe.

"As the system evolves, the strong gravitational pull that embryos receive from the giant planets and from each other deforms the embryos' orbits, which begin to cross. A cascade of collisions follows, forming planets as the embryos merge and collect planetesimals. Leftover planetesimals become asteroids." (John Hartnet)

Mystery #9:
Black holes

One of the least explained phenomena in space are black holes. Among other things, they are supposed to explain the unexplainable. Some scientists theorize that they are created when a space object becomes "large enough" to "collapse" into something they can't explain but still could "absorb" nearby matter.

Incidentally, no such thing as a black hole was ever detected. Some of the suspected black holes are located very close to stars without "swallowing" them.

Supposedly, a black hole is a region of spacetime that is extremely distorted by the presence of a sufficiently dense mass. The gravity in this region appears to be so strong that anything getting too close to the black hole will never escape.

It is believed, some black holes have masses comparable to that of our sun. According to one theory, such stellar-mass black holes are thought to form when a star consumes the last of its nuclear fuel and collapses under the influence of its own gravity.

Different supermassive black holes are believed to have masses millions or even billions of times greater than this. These supermassive black holes seem to be present at the centers of most, if not all, large galaxies such as the Milky Way.

One particular black hole is called "supergiant black hole" and is assumed to be billions of times bigger than our sun. For the first time, cosmologists were able to take a picture of what they claim is a black hole. This appears to be impressive and coincides with Einstein's relativity theory. One of the many problems is that they are so huge and so far away from us that scientists can't imagine how this could have been done during the time frame calculated.

Even though black holes are supposed to be the final "proof" for the Big Bang theory, it poses more questions than answers. Is this science or fiction? The future will tell.

Mystery #10:
Dark matter

The probably most mysterious mystery of the universe is the dark matter. It is supposed to comprise 85 percent of the universe. Nobody can see it, but the existence of dark matter is absolutely essential for the survival of the Big Bang model. Therefore, secular scientists insist it must be upheld at any cost.

This is how NASA theorizes about the phenomenon: "For the first 150 million years after the Big Bang, there were no galaxies or stars or planets. The universe was featureless. As time passed, the first stars formed. Stars collected into galaxies. Galaxies began to cluster together. Those clusters are made up of the galaxies and all the material between the galaxies. Clumps of matter in smashed into each other and the planets in our solar system began to form around the sun. Something must hold our solar system, galaxies, and clusters of galaxies together. And gravity is that glue." In some clusters, the space between galaxies is filled with gas so hot, scientists cannot see it using visible light telescopes. The gas-only can be seen as X-rays or gamma rays. Scientists look at that gas and measure how much there is between galaxies in clusters. By doing this, they discovered that there must be five times more material in the clusters than we can detect. The invisible matter that we can't detect is called dark matter."

What do we know for sure dark matter cannot be:
- Failed stars
- Clouds of gas

- Dust grains
- Asteroids or comets
- Small clumps of normal matter
- Ionized plasma
- Black holes

On the other side, we assume the following assumed characteristics of the dark matter:

- Should be very cold.
- Doesn't interact with itself, light, or normal matter.
- Appears to have a dominant effect in a galaxy
- Causes a gravitational effect at places with no matter.

Most secular cosmologists believe in dark matter mainly because it underscores their worldview opposing the Biblical creation. The main problem is that the dark matter is not visible. Scientists are trying to overcome this problem through alternative technologies without any solid results beyond new and old theories.

"Although this is true, dark matter has nevertheless become extremely important to secular scientists. Big Bang cosmologists now invoke dark matter in their theories of galaxy formation. Furthermore, Big Bang cosmology demands that this dark matter be composed of exotic, never-before-detected particles. This is because, in order for the Big Bang to produce the observed abundances of helium and hydrogen, Big Bang cosmologists must assume a particular value for an adjustable parameter in their model. Once this is done, the Big Bang model can only generate enough "normal" matter (atoms) to account for about 15% of the matter thought to exist. This means that, by their reckoning, the dark matter can't be made of atoms. But they have ruled out other forms of non-atomic matter (neutrinos, free electrons, etc.) for other reasons. This means that some exotic, new form of matter is the only remaining option." (Jake Herbert, Ph.D.)

The ongoing mystery of what dark matter actually endures the ultimate challenge of modern fundamental physics. The central question is whether it is indeed a missing mass source, such as a new type of matter, or whether the gravitational law is simply different at gigantic length scales, physicists speculate.

Still, new "discoveries" are published about some specific aspects all the time. They are supposed to support their theory and bring them closer to the badly needed proof. So-far, dark matter problems remain dark.

Mystery #11:
The origin of the earth

For Wikipedia, the geological origin of the earth is a scientific fact. "The geological history of earth follows the major events in earth's past based on the geological time scale, a system of chronological measurement based on the study of the planet's rock layers. Earth formed about 4.54 billion years ago by accretion from the solar nebula, a disk-shaped mass of dust and gas left over from the formation of the Sun, which also created the rest of the solar system. Earth was initially molten due to extreme volcanism and frequent collisions with other bodies. Eventually, the outer layer of the planet cooled to form a solid crust when water began accumulating in the atmosphere. the moon formed soon afterward, possibly as a result of the impact of a planetoid with the earth. Outgassing and volcanic activity produced the primordial atmosphere. Condensing water vapor, augmented by ice delivered from comets, produced the oceans."

Basically, this is what we are being taught in schools and at universities worldwide. For secularists, this is pretty much written in stone. They hate to discuss their theory because they claim it is "scientific" and "final." Fortunately, no science result is ever really final. Science evolves by observation, and every new day may bring new facts, viewpoints, and models.

A brilliant scientist with the name Robert V. Gentry compiled breathtaking discoveries totally refuting the evolutionist dogma. Over five decades of research, strictly adhering to the laws of natural sciences, resulted in facts nobody can dispute anymore.

The radioactive granite

The major basement rock on earth is granite. Many geologists argue that they originate from the slow cooling of molten lava to reach a solid form. Typical for granite is that it contains large amounts of the radioactive polonium, which automatically decays in less than three minutes. According to the evolution theory, the earth was molten for "millions of years."

If this theory is true, no polonium halos should be left. In reality, today's granite in earth holds trillions of polonium halos. This means that the major rock formation of the earth came into existence in less than three minutes. The only logical explanation, according to the law of natural sciences, is that the earth was created in a very short time - and not in billions of years.

The granite from molten lava

Experiments show that the granite with large crystals never can be made from molten lava rock, which always produces ryolite with much smaller crystals - but never granite.

Mystery #12:
The age of the earth

The age of the earth is one of the hottest topics discussed in science, history, philosophy, and theology for thousands of years. There are two major competing theories around.

- The young-earth theory was established by creationists.
- The old earth theory was proposed by secular evolutionists.

The National Geographic summarizes the prevailing secular wisdom as follows:

"Earth is estimated to be 4.54 billion years old, plus or minus about 50 million years. Scientists have scoured the Earth searching for the oldest rocks to radiometrically date. In northwestern Canada, they discovered rocks about 4.03 billion years old. Then, in Australia, they discovered minerals about 4.3 billion years old. Researchers know that rocks are continuously recycling, due to the rock cycle, so they continued to search for data elsewhere. Since it is thought the bodies in the solar system may have formed at similar times, scientists analyzed moon rocks collected during the moon landing and even meteorites that have crash-landed on Earth. Both of these materials dated to between 4.4 and 4.5 billion years."

Considering the fact that the radiometric age test is extremely unreliable (more about this later in this book), the old earth theory is mostly based on claims and theories. The young-earth theorists, however, counter them mostly with logics, physics, chemistry, and geology.

The slowing earth rotation
A very interesting fact is the rotation of the earth, which happens at a speed of about 1,000 miles per hour. Evidently, this speed is gradually slowing down. If the earth were really billions of years old, the rotation would have stopped millions of years ago.

The decaying magnetic field of the earth
A very similar occurrence is evident in the magnetic field of the earth. Unfortunately, it is decaying, too. Since 1835, global magnetism has decreased by 14 percent. According to calculations based on this result, just 7,000 years ago, the magnetic field on earth would have been about 32 times stronger than today. About 20,000 years ago, this force would have liquified the earth totally. One million years ago, earth magnetism could have vaporized the universe.

The escaping natural gas
Our natural gas and oil reserves are mostly located under a porous rock such as sandstone, often sealed at the top by another firm rock formation. Naturally, some of the gas can still escape through a shale cap. However, if this should have happened for millions of billions of years, there should be nothing left.

The oil pressure
A very similar effect has been observed measuring the oil pressure, which is much too high considering millions of billions of years. If the evolution theory

were true, the oil pressure should have bled-off a very long time ago.

Accordingly, the current oil seepage in oceans is not the effect of organic oil pollution. If the oil was seeping out of the floor for billions of years, there should no more offshore wells existing anywhere. Also, there should no ancient oil reservoirs to be found.

Why is the earth's center still molten hot?

We know for a fact that the inside of the earth is still incredibly hot, consisting of molten rock. How is this possible after billions of years?

At the same time, why do we still have big and violent volcanic eruptions all over the world? We still can count thousands of active volcanoes. After billions of years, no planet can retain the heat we have.

The zircon-lead and helium ratios

It has been proven that radiogenic lead leaks out of zircon at very high temperatures. Incidentally, virtually no lead diffusion can have happened out of lava-based zircon on earth.

Very similarly, helium can escape out of crystals rapidly at extremely high temperatures such as in the center of the earth. All helium in crystals should disappear by now if the earth were older than thousands of years.

The soil-water ratio
Interestingly, geologists consider the earth's soil to be still pretty much water-soaked but slowly drying out. This effect appears to be explainable through the big flood but not by the very old earth.

Earth's topsoil usually measures about eight inches. Geologists calculate that it takes 300 to 1,000 years to build one inch. According to this reckoning, the age of the earth should not exceed a maximum of ten thousand years.

The Niagara Falls
The stream of Niagara Falls appears to erode the cliff at a rate of approximately 3.5 feet per year. Since its length is about seven miles long, this would point to an age of 5,000 to 10,000 years.

The river path
The Mississippi River dumps close to three hundred million cubic yards of soil into the Gulf of Mexico each year, which is enlarging Louisiana slightly all the time. Careful calculations, however, point out that this effect hasn't happened much longer than about 4,000 years.

The ocean floor
The oceans are full of all kinds of living organisms. When they die, they drop to the ocean floor and build a mud-like ocean floor. This is supposedly being done at a rate of one inch every 1,500 years. Actual measurements prove that the ocean floor ooze is not very thick and can't relate to an old age theory of the earth.

Scientists estimate that about thirty billion tons of sediments are added to the ocean floor each year. Billions of years would produce approximately sixty to hundred miles thick ocean floor by virtually eroding away all continents. The actual thickness is measured at a half-mile.

If the erosion of the oceans occurred for billions of years, why aren't underwater cliffs, mountains, and valleys fully filled with sediments yet?

The Sequoias tell their story
The giant sequoia trees in the Sierra Nevada mountains belong to the most unique plants. They are not just the largest trees on earth, they are in some ways eternal. They virtually never die, and some of them are up to 4,000 years old, according to their tree rings. No insect or forest fires can destroy them, only humans with their saws.

What does this discovery tell us about the age of the earth? It shows us that there is only the first generation of Sequoia gigantea trees we can find. How can evolutionists explain that?

What historical records reveal
The prevailing "knowledge" in our schools and colleges claims unmistakably that our earth is billions of years old. How come that all reliable historical records we can find start around 3,500 B.C.?

Among the oldest historical accounts are documentations of early Egyptian and Hebrew civilizations. Some confusion with Egyptian dating lasted for a century and has been lately corrected from 5,800 B.C. to approximately 2,900 B.C.

Due to incorrect radiocarbon dating methods (to be explained later in detail), the first Sumerian records in Babylon had to be also dramatically corrected.

The Genesis records show that the creation occurred about 4,000 B.C. and the worldwide flood approximately 2,348 B.C. The first books of the Bible were written between 1510 and 1450 B.C.

The solar eclipses
Astronomical data is much more reliable than the most used radiometric age tests. Very telling are records of the solar eclipse. Interestingly, there are no records of a solar eclipse older than 2,250 B.C.

The written records
Written historical records are always extremely important in judging the past. The oldest Sumerian pictographs are dated around 3,500 B.C. The first Western-type of scriptures appears to have happened around 1,550 B.C. Importantly, no doubtless verified finds of any civilization can be discovered beyond 3,500 B.C.

The languages spoken
All intelligent human communication is expressed by their language. The oldest record probably is the Indo-European language family. Why can't we find any reliable records of a spoken word beyond 3,000 B.C.?

Are star clusters moving away?
We observe lots of star clusters in the universe. They move very rapidly in a specific direction. How can still stay together if they do that for billions of years?

Not enough hydrogen
Some stars are so huge that they radiate their energy up to one million times more than our sun. How can they sustain their hydrogen reserve for billions of years?

Why is hydrogen still left?
Hydrogen is permanently being converted into helium to make stars shine. Contemplating billions of years, there should be any hydrogen left. How can this be explained?

The shrinking sun problem
Researchers discovered that the sun is constantly shrinking at a steady rate of arcseconds per century. Assuming this rate of shrinking, 50,000 years ago the sun would be so hot that our oceans would boil. In a much shorter time than that, any kind of life on earth would be impossible.

Sun emitting solar neutrinos
An old sun should emit solar neutrinos. Since this is not the case, the sun can't be very old.

Comets circling the sun
Comets circling the sun are comprised of rocky debris, frozen gas, and water. Flying around the sun, they are slowly disintegrating. Any billions of years old comet should have self-destructed and disappeared a long time ago.

Where is the water?

The latest research indicates that comets mostly contain water. They are striking the earth all the time. If this happened for billions of years, the earth should be overfilled with water.

The solar wind radiation

The solar wind is the radiation of the sun flowing outward. Researchers still detect big amounts of these particles orbiting the sun. According to a mathematical calculation, all particles smaller than 1/100,000 of a centimeter should have been blown out a long time ago if the solar system is really billions of years old.

The Saturn ring phenomenon

The Saturn ring is mostly composed of solid ammonia, which should vaporize into space after billions of years. Also, according to astrophysical research, the permanent bombardment of Saturn rings could not survive any longer than only 20,000 years.

The Galilean moons

Io, the third largest and innermost of the four Galilean moons of Jupiter, has over sixty drastically active volcanoes ejecting plumes at the speed of 2,000 miles per hour and up to 160 miles into space.

Being one of the smallest moons, it would be technically impossible to maintain the heat and to spew gases for billions of years. Also, the moons are so different from each other that it appears very unlikely to share the same origin some billions of years ago.

The radioactive decay
The decay of radioactive uranium and thorium generates helium. Assuming the decay rate for billions of years, the earth's atmosphere should have a dramatically larger quantity of helium present.

The carbon 14 disintegration factor
The carbon 14 disintegration shows a similar picture. The current amount of radiocarbon in the atmosphere coincides with the quantity created in just thousands, not millions or billions of years.

Where are the meteorites?
About 20 million meteorites collide with the atmosphere of the earth every twenty-four hours. Multiplying this action by millions or billions of years would amount to thousands of times higher collection of meteors and meteor dust on earth than we can actually observe.

Meteor craters at earth's surface
We can find meteor craters all over the world. They are usually easy to locate and are quite big. Incidentally, they always lie very close to the surface. This proves that they are all very young, just thousands of years. No deep craters can be found in lower rocks indicating that there are no millions of billions of years old meteor hits on earth. Also, no meteorites can be found in deep earth strata.

Mystery #13:

The moon

Our moon hides a number of mysteries our space expeditions tried to uncover. One of them is the origin and age of the moon. The prevailing theory is that the moon is as old as the earth, billions of years old. Can this be verified?

The moon dust

The moon is constantly collecting dust. Considering the annual rate of dust gathering and the suggested age of 4.5 billion years, the current dust layer should be up to sixty miles high. The actual dust cover is just about two to three inches high, which mathematically amounts to 6,000 to 8,000 year.

The lunar soil

The lunar soil analysis does not indicate any required soil mixing normally produced by billions of years old moon.

Uranium and thorium

The analysis of moon rocks revealed uranium 236 and thorium .230 in the stones. Both elements are short-lived. If the moon was only 50,000 years old, both isotopes would have decayed into a lead long time ago.

The moon rock

Similarly, moon rocks were also tested for their radioactivity. The radiometric test methods produced some conflicting results, but they clearly showed that they can't be very old.

The lunar gases
The research of the lunar gases also generated some interesting results. Taking into account the intensity of solar winds, the amount of solar gases on the moon does not exceed a maximum of 10,000 years.

The still thermally active moon
The moon still shows some lunar activities such as quakes, lava flows, and gas emissions. It is not at the thermal equilibrium yet. At the proposed age of billions of years, the moon should be very cold without any thermal activity.

The moon moving away
The moon is slowly but gradually moving away from the earth. This confirms that the moon used to be much closer than it is today. Just 20,000 to 30,000 years ago, according to mathematical calculations, the moon would have been so close that it would have crashed into the earth. According to the lunar recession, the moon must be younger than about 10,000 years.

Mystery #14:
The age dating methods

As "scientific" most of the age dating methods may sound, they are often very problematic and highly inaccurate. In short, most of the current dating results are simply wrong, often by a factor of ten or even much more.

There are several radio-dating testing methods delivering very different results and requiring a lot of skepticism.

- Uranium-thorium-lead dating indicating the disintegration of uranium and thorium into elements such as radium and helium, which finally results in the lead.
- Rubidium-strontium dating, which measures the decay of rubidium into strontium.
- Potassium-argon dating, which is working by measuring the conversion of potassium into argon and calcium.
- Carbon-14 radiocarbon dating relies on the formation of radioactive elements of carbon in the atmosphere and their final decay into a chemically stable carbon isotope.

All of these test methods are extremely problematic because they are all based on mostly unreliable assumptions. All of the assumptions must be fully correct to secure the expected accuracy.

In reality, nobody can guarantee the exactness of them and they are, as the word says, just assumptions. Therefore, most of the test results are only estimates no scientist can honestly rely upon.

- Every testing method must be a completely closed system with no contamination of parent or daughter substance during the decay process.
- None of the testing procedures may contain any of the daughter products.
- The decay and the process rate must stay the same and can't ever be changed.
- Any change in atmospheric conditions around the earth could affect the clocks in radioactive minerals.
- Any change in the Van Allen belt can strongly change the transformation of radioactive materials.
- The radioactive dating is only correct if the clock can start at the beginning and no daughter products are present.

Dozens of different dating methods are used with greatly different results. Very popular is the Carbon-14-dating method. Fewer than fifty percent of these results are considered somehow "acceptable." Most of them are tossed out right away because they are not consistent and are therefore unreliable.

As Vance Farrell explains in his highly recommendable publication called "The Evolution Handbook," the commonly used radio-dating has a series of major problems turning this method virtually worthless.

- Inconsistency in the type of carbon.
- Variations between different samples.

- Loss of Carbon-14 through rainfall and other factors.
- Deviations in the carbon found in the atmosphere.
- Sunspot effect on Carbon-14.
- Radiocarbon date comparison showing often results in millions or even billions of years younger than tested.
- Neutrino radiation in the atmosphere can change and influence the radiocarbon levels.
- All kinds of cosmic rays influence the atmosphere and test results.
- The weakening magnetic field of the earth may influence the testing.
- The moisture and other atmospheric changes affect the C-14 quantities.
- Earlier more water and warmer temperatures affect the results also.
- Dramatic changes after the big flood lead to atmospheric and other climate changes and therefore influence the test results.
- Even the latest tests appear to be inaccurate. There are many examples of terribly wrong dating. Freshly killed seals showed an age of 1,300 years. Freshly cut pieces of a tree produced an age of 10,000 years, etc.
- The developer of the radiocarbon dating, Willard F. Libby, found a series of discrepancies, assuming the build-up of the terrestrial radiocarbon was inaccurate.
- Younger dates after 600 B.C. appear to be more accurate.

Summarizing, the historical radiocarbon dating is very inaccurate. "The troubles of the radiocarbon dating method are undeniably deep and serious..." (Anthropological Journal of Canada)

Unfortunately, other age dating methods, suggesting the earth is millions or billions of years old, are also less than reliable. One of them is the amino-acid-dating and is based on the decomposition of amino acids, the building block of proteins.

The main drawback of this test is that some of the twenty amino acids decompose much faster than others. Therefore, estimates must be made when an examined animal died. How? Of course, estimates are estimates, far from being reliable or correct. Still, according to the current status of the research, no really "ancient" examples have been found anyway.

There is a series of other dating methods such as astronomical dating or stalactite dating in use. Vance Ferrell describes this in his book in great detail.

Mystery #15:
The background radiation

Astrophysicists can prove that there is very weak radiation of heat (just 2.73 degrees above the absolute zero) present in outer space. Proponents of the evolution theory claim this is the single best proof that the Big Bang happened. Hypothetically, this is the "remnant of heat caused by the Big Bang explosion" billions of years ago.

However, this occurrence would have to fit four conditions to be considered a validation of the evolution theory:

- The heat must come from one direction, the alleged Big Bang explosion.
- The heat must have the exact radiation strength to match the Big Bang mathematically.
- The heat must have an exactly defined spectrum.
- The heat must not be smoothly distributed.

However, the actual analysis shows a different picture:

The heat distribution comes from all sides
The heat distribution is omnidirectional meaning that it cannot come from one source such as the Big Bang explosion.

The heat is too weak
The heat is too weak. Theoretically, it should be much hotter. An explosion would cause a much stronger occurrence.

The heat shows a different spectrum
The heat does not show the same spectrum as required by physics. What does this mean?

The heat distribution is too smooth
The heat appearance is substantially too smooth to be generated by an explosion.

Mystery #16:
The "redshift" method

Determining the age of solar objects is particularly important and fascinating. The discovery of the redshift method appeared to be the desired breakthrough, at least it was celebrated as such. How does it work and how reliable is it?

We all know that the "white" light can be split into the rainbow with a triangular prism. Using a spectrometer, astrophysicists analyze the light of stars to determine the color and classification of the stars.

The ultraviolet light is on one side of the spectrum and has a higher frequency and a shorter wavelength than the visible blue light. Infrared, however, is on the other side of the visible range. Astronomers call it "red."

The theory says the farther a star or a galaxy is from our standpoint the more the measurable light spectrum is shifted towards the red. This is called redshift.

In theory, this is supposed to prove that the universe is expanding outward from the Big Bang position. Hypothetically, this means that objects moving torch us appear bluer, and the stars moving away from us become redder and redder. This is named the "redshift" theory.

It appears obvious that the redshift shows some correlation to the distance of a star investigated. There are four different explanations offered by different scientists. Vance Farrell offers more detailed information about each version.

- The speed redshift
- The gravitational redshift
- The second-order Doppler shift
- The energy-loss shift

The believer in the Big Bang theory, however, claims that the speed is the only cause for the redshift and it documents the expanding of the universe away from the explosion site.

Let's fact-check this theory. First of all, almost all galaxies are red-shifted. What does this mean?

Objects not moving away
If only the speed is the cause for the red-shift, this would mean that everything in space is moving away from the earth and not from a potential place of a Big Bang. This is impossible. The only logical explanation would be that everything in space is moving away from the actual place of the purported explosion.

The stars moving toward earth
In fact, some of the closest stars are moving toward the earth but are still redshifted. How come?

The contradicting quasars
As Vance Farrell points out, quasars strongly disapprove of the speed theory of the redshift. Some of them travel eight times faster than the speed of light. Considering the speed theory, they should be at an impossibly great distance from us.

The factor light

The light has weight and can be pulled by gravity and change the redshift result.

No blue-shifted stellar light

We never observed a blue-shifted stellar light spectrum, which conflicts with the speed theory.

Mystery #17:
Fossils and strata

Fossils are featured in most national parks, textbooks, and museums around the world. The mystery is, were they formed over millions of years, as evolutionists claim?

First of all, fossils don't need millions of years to form. In a simple laboratory setting, this process can be accomplished fairly quickly. Under specific conditions, some fossils have been formed in just a few days.

Fossils are the mineralized remains of once-living organisms. This can be plant or animal bodies, partially or completely replaced by minerals. Sometimes, they are just impressions that show the shape of a creature, or tracks left behind by a traveler, or other remnants that testify to the lives of long-gone organisms.

Surprisingly, some occasional fossils even have original organic soft tissue encased in rock, showing that their organisms couldn't have been dead for millions of years, as Charles Darwin claimed.

Naturally, all fossilized creatures appear suddenly and are fully formed in the rock record. There is no clear history of evolutionary transitions. By the way, this is consistent with the Bible's assertion that God created life forms to reproduce within their own kinds.

In order to get a perfect depiction, fossils must have formed very quickly, before the animal or plant completely decayed or was scavenged. Logically, that documents that fossils formed through sudden, catastrophic circumstances.

According to the creation theory, most fossils can be attributed to the worldwide catastrophe of Noah's Flood, in which countless creatures were killed and then rapidly buried, or to its residual effects.

Whatever happened at that time, fossils are the undeniable proof of the process at that time. Facts don't change, just the interpretation of them.

According to evolution models for fossil records, they are supposed to show:

- A comprehensive change of organisms through time.
- The primitive organisms gave rise to complex organisms.
- A gradual derivation of new organisms producing transitional forms.

However, the fossil records found don't confirm that at all.

One of the most obvious examples is trilobites. They appear suddenly in the fossil record without any kind of transitions. That means that there are no fossils between simple single-cell organisms, such as bacteria, and complex invertebrates, such as trilobites.

Furthermore, fish also have no ancestors or transitional forms to show how invertebrates, with their skeletons on the outside, became vertebrates with their skeletons inside.

What we found are only fossils of a wide variety of flying and crawling insects appear without any form of transitions. Dragonflies, among many examples, appear suddenly in the fossil record. The highly complex systems, enabling dragonfly's aerodynamic abilities, have no ancestors in the fossil record at all.

In the entire worldwide fossil record, there is not a single unequivocal transitional form proving a causal relationship between any two species. From the billions of fossils discovered, there should be thousands of clear examples if they really existed. This is how Vance Ferrell describes the problem:

- There are no transitional species preceding or leading up to the first multi-celled creatures that appear in the Cambrian, the lowest stratum level.
- There are no transitional species elsewhere in the fossil record.
- The species that appear in the fossils are frequently found in many different strata.
- The great majority of the species found in the fossils are alive today.

Considering all the fossils discovered, we find no transitions from one kind of creature to another. Only individual, distinctive plant or animal kinds are discovered.

- Different kinds of animals appear abruptly and fully functional in the strata, with no proof of ancestors. "Evolution requires intermediate forms between species and paleontology does not provide them" (David Kitts, paleontologist, and evolutionist).

- Evolutionists admit that there is no proof of evolutionary history for even one group of modern plants. "I still think that to the unprejudiced, the fossil record of plants is in favor of special creation." (Evolutionist Edred J.H. Corner)
- The history of evolution is supposed to be filled with temporary, intermediate stages of evolution, from amoeba to humans. However, most fossils are very similar and often identical to creatures living today.
- The lack of evidence for evolution continues despite millions of fossils already discovered and the sediments already explored.

The Institute for Creation Research collected a series of specific examples of how fossils defy Darwin's evolution theory.

- Fossil anemone footprints, found in Newfoundland, are supposed to confirm the evolution theory. They should be 565 million years old. Unfortunately, their muscle composition defies evolution's own theory because this kind of muscle development should have happened millions of years later.
- A supposedly 34 million years old fossil cuttlefish still has some of the original tissue left. But the fact that it has not yet completely decayed is only expected if the cuttlebone was fossilized just thousands of years ago.
- Researchers investigated a fossil of a bird feather, which was dated as millions of years old. Surprisingly, the analysis determined the color of the feather. "They found trace-metals

that have been shown to be associated with pigment and organic sulfur compounds that could only have come from the animal's original feathers," according to the University of Manchester. Incidentally, any organic molecules like pigments, and especially proteins, shouldn't be there if the specimen is older than a million years. The same kind of discovery was made by the examination of a fossil moth.

Fossils are a great embarrassment to the evolutionary theory and offer strong support for the concept of creation." (Dr. Gary Parker, biologist, paleontologist, and former evolutionist)

Strata layers
The major formations of the earth's crust are sedimentary rock beds. These shapes were formed by rapid erosion, transportation, and deposition by water. There is no global evidence of long periods of time between these layers or indications that these layers took long periods to form.

Strata are called layers of earth sediments deposited all-over the world, according to the evolution theory, for billions of years. Interestingly, sandstone is a major feature of the lower part of the Grand Canyon. However, the very same rock layer is found in Utah, Wyoming, Montana, Colorado, South Dakota, the Midwest, the Ozarks, and even in northern New York state. Corresponding formations are found across wide portions of Canada, eastern Greenland, and Scotland.

What does this mean?

The evolution proponents tell us, the Tapeats sandstone at the bottom of the Grand Canyon is about 550 million years old, while the Kaibab Limestone at the top is only 200 million years old.

Astoundingly, these sediments were uplifted to their present high elevation, up to 7000 feet at the rim, about seventy million years ago, meaning the Tapeats was already 480 million years old at the time of uplift and "deformation." Do they want us to believe that this kind of rock was bend and uplifted millions of years after it turned into hard stone?

The only logical explanation, in accordance with the law of physics and geology, was provided by the Institute for Creation Research scientist John D. Morris, Ph.D.: "In Grand Canyon Park, most of these sediments, which were laid down horizontally underwater, remained horizontal after uplift. But the uplift Beverly deformed these same sediments along the flanks of the plateau, in some areas leaving them in a vertical orientation. In my favorite spot, the Tapeats, which today is an extremely hard rock, was bent from horizontal to vertical in a space of 100 feet or so. The nature of this deformation shows that the sediments were almost certainly still soft when bent. They had not yet had time to turn hard. But it only takes a few hundred years at best for sandy sediments to turn to sandstone in the presence of high overburden pressure and adequate cement. Therefore, we are justified in concluding that the Tapeats was not 480 million years old at the time of uplift. It all happened in a short period of time, while the sediments were still soft."

Most geologists nowadays accept that catastrophic processes are necessary for the deposition of nearly all rock types. This means, only a short time was needed to form each bed of sediments, which eventually hardened into sedimentary rock. The central question is how much time elapsed between the deposition of one bed and the deposition of a directly overlying bed.

A very commonly found feature, seen in many rock layers in many locations, is the presence of ripple marks, which form as water moves over a surface the sediments. Almost all sedimentary rock layer was deposited underwater.

Under this condition, the sediment will soon erode and be washed away, especially in soft, unconsolidated sediments. Even on a hard rock surface, markings will erode in a few decades. We cannot estimate with certainty how much time went by during this process, but we can determine with confidence that this process did not take millions or billions of years as claimed by evolutionists.

We cannot examine the fossils, strata layers, and sediments without taking the catastrophic event of gigantic proportions into account that twisted the rocks, hurled mountains upward, pulled water out of the earth, caused thousands of volcanoes to erupt, and even affected the atmosphere dramatically.

The Big Flood theory
The geological formation of the entire earth shows that this couldn't have been a local or regional event.

The only logical and scientifically feasible explanation is the Big Flood as described in the Bible.

Interestingly, the granite shows no fossils. This means that it was formed before the flood. Within the strata, however, billions of fossils were found. Many of them are in such a flawless condition confirming that the animals of plants found were buried very suddenly under great pressure. Any slow process would have caused a natural decomposition before turning into a fossil.

The Flood theory confirms what physics, geology, and other sciences would expect to happen at the Big Flood.

- Animals living at the lowest levels are buried at the lowest strata.
- Different creatures buried together to reflect their living presence in the same ecological sphere.
- The hydrological and drag forces tend to "sort" the creatures in accordance with the laws of physics.
- Animals without backbones living at the sea bottom are found at the lowest level of the strata.
- Fish are found at the surface.
- Amphibians and reptiles are found above the fish but under land animals.
- Just a few plants or land animals are seen at lower strata levels.
- Land plants are discovered next to amphibians.

- Mammals and birds are located higher than amphibians and reptiles.
- Some animals tending to stick together in the case of danger are seen close together.
- Large, strong animals separate themselves from the small ones.
- Relatively few birds are found in the strata because they were able to fly to the highest points in the area.
- Very few humans are found in the strata. Otherwise, they were found only at the highest levels.

Independently from the Christian history, a survey among 120 tribal groups in North, Central, and South America disclosed their ancestral knowledge of the Big Flood dozens of times.

- God punished general wickedness on mankind.
- God considered the flood necessary.
- One good family with eight members was protected.
- A giant boat was constructed.
- The family, animals, and birds entered the boat.
- The flood flooded the rest of the population.
- The flood covered the entire earth.
- The boat arrived in a high mountainous area.
- A few birds were sent out first.
- The people left the boat along with the animals.
- The survivors worshipped God for saving them.

- After reaching the land, God spoke favorably to them.

"There are many descriptions of the remarkable event [the Genesis Flood]. Some of these have come from Greek historians, some from the Babylonian records; others from the cuneiform tablets [of Mesopotamia], and still others from the mythology and traditions of different nations, so that we may say that no event has occurred either in ancient or modern times about which there is better evidence or more numerous records, than this very one. It is one of the events which seems to be familiar to the most distant nations - in Australia, in India, in China, in Scandinavia, and in the various parts of America." (Stephen D. Peet, "Story of the Deluge")

Mystery #18:
Natural selection

The basic teaching of Darwinism is "natural selection." This means, when a plant or animal produces an offspring, variations appear. Some of the offspring will naturally be slightly different from other offspring.

Darwin declared that these variations, he called "natural selection," have generated all life forms on our planet. From the simplest bacteria to pine trees, jackals, clams, zebras, frogs, grass, horses, and even humans. Quite a claim!

"So far as we know, natural selection is the only effective agency of evolution."(Sir Julian Huxley, Evolution in Action)

"Natural selection allows the successes, but 'rubs out' the failures. Thus, selection creates a complex order, without the need for a designing mind. All of the fancy arguments about a number of improbabilities, having to be swallowed at one gulp, are irrelevant. Selection makes the improbable actual." (Michael Ruse, Darwinism Defended)

Evolutionists not only claim that the natural selection produced everything, but they also say that everything happened completely "randomly," by chance. This in reality means, nothing was really "selected." Everything was accidental, what happened, just happened. This is the opposite of any "selection." Even the name of the proposed process is misleading.

Let's examine the terminology:

- Natural selection can only be considered if a plant or animal crosses the species barrier and creates a new species. Changes within a species, such as breeding a different kind of dog, can't be considered evolution. There are hundreds of different kinds of dogs, but they are still dogs. This is only a variation.
- Species are called the fundamental type of a living organism such as fish, bird, or a cat.
- Variations mean mostly slight changes within a species such as breeding a new form of a cat such as a white, black, or yellow cat.
- Mutational changes are displayed when a DNA defect occurs and the offspring changes some characteristics such as the shade of the fur.
- The survival of the fittest is the central argument for the evolution theory. It is supposed to exhibit that accidental mutations produce "misfit" variations. Just by accident, only the "fittest" survive, generated by a genetic mistake, and this is called natural selection. Even if accepted as claimed, this is not evolution but at a variation within a species at best.

One of the most frequently suggested "proofs" of evolution is the peppered moth. At the time of Darwin, they were all light-colored with occasional dark spots. For centuries, some of the moths started to have darker wings.

Around the 1950s, 98% of them were dark, but some were still light. Evolutionists cite this phenomenon as their "proof of evolution." Since all of them still were moths, remaining within one species, they were just varieties of the species. Of course, this is not evolution, but simply a color change back and forth within a stable species.

Similar observations have been made on bacteria, flies, and pigeons who produce an astonishing variety of shapes and colors. Also, apples, grapes, and Galapagos finches, just to name a few, impress with an amazing variety but still remain the very same species.

"There is no agreement on the extent to which metabolism could develop independently of genetic material. In my opinion, there is no basis in known chemistry for the belief that long sequences of reactions can organize spontaneously - and every reason to believe that they cannot. The problem of achieving sufficient specificity, whether in an aqueous solution or on the surface of a mineral, is so severe that the chance of closing a cycle of reactions as complex as the reverse citric acid cycle, for example, is negligible. The same, I believe, is true for simpler cycles involving small molecules that might be relevant to the origins of life and also for peptide-based cycles." (Lee M. Spetner)

Mystery #19:
The DNA code

For every life to exist, an extremely complicated information system is needed to create and regulate life functions. This data system must also be able to accurately copy itself to produce the next generation. DNA and RNA represent the essential information system of life.

One of the most important discoveries of the last century was the discovery of the DNA molecule. DNA, or deoxyribonucleic acid, is the genetic code, or blueprint, that plays the central part in defining who you are. In short, it contains all the genetic information about you. This means really everything, from the color of your skin to your gender, size, talents - to your intellect and most distinctive personality features.

Every cell in our body contains a copy of your DNA, our genetic code. It is called the gene pool of genetic traits. It is also named the genome. Considering the huge amount of information DNA contains, this code within each DNA cell is more complicated than anything we know.

Scientists calculated that all information of the world would fit in a teaspoon full of DNA.

No life can be created or replicated without DNA. Can anybody imagine that such a complicated and powerful system of life could be created by a random accident from dirt and water, as Darwin and his followers want us to believe?

Not surprisingly, no scientist was ever able to create a DNA molecule in a laboratory. If they, working in multimillion-dollar equipped and stocked labs, cannot make DNA and RNA, how can a random action of sand and dirty water produce it millions or billions of years ago?

"What does this mean? Simply, in order to make the DNA, you have to have DNA in the first place! You have to have the DNA code within the cell before you can duplicate a DNA code. Without the complete code in the first place, there is no way to make the code necessary for every living cell!"
(Lawrence O. Richards)

Mystery #20:
The humans and the apes

There is a widespread belief that humans are descendants of apes. This is what Charles Darwin's evolutionists claim and children learn in our schools.

We all have seen Darwin's deceitful drawings showing a line of monkeys turning into a human creature. At the first glance, it appears reasonable, but any closer comparison of humans with orangutans or other apes proves the deception on many levels.

First of all, there are a few very obvious differences:

- Different birth weight
- Different teeth eruption
- Different neck position
- Different (straight) walk

Remains of cavemen, such as Neandertals excavated in Germany, were supposed to prove the evolutionary link between monkeys and modern humans. The bones of the Neandertal man were carefully examined by scientists and they concluded that the bones are virtually identical to today's human bones.

The slight differences, evolutionists hoped to prove human evolution, were just effects of their arthritis. Of course, the bones were not millions of years but only several hundred years old. Many such bone pieces were found around the world such as the Peking man, the Rhodesian man, the Nebraska man, and more.

Additional questions arose nobody is able to answer:

- We always found only one specimen. Why?

- We always found little pieces, never a complete skeleton. Why?
- Why did the bones not decay if found in humid regions and were supposedly millions of years old?

Summarizing the findings, the evidence allows only one conclusion. Mankind did not "evolve" from a lower form of life. Many anthropologists still maintain that mankind descended from an unknown ancestor, which poses even more unanswered questions:

- Why do humans have a different number of vertebrae than apes?
- Why are our cranial nerves, emerging directly from our brain, totally different?
- Why is our DNA dramatically different from apes and all other animals in wildlife?

With the exception of some general body similarities, humans have virtually nothing biologically in common with apes.

Mystery #21:

The unanswered questions

Over one hundred fifty years after the introduction of the evolution theory by Charles Darwin, none of the essential questions have ever been answered.

The simple reason is that Darwinism is incompatible with logic, history, creation, and in particular with natural laws of physics, chemistry, cosmology, geology, and many more.

Unbiased scientists, relying on natural laws, and Christian organizations such as the Institute for Creation Research, and Creation.com want countless questions answered before evolutionists continue to maintain their claim that evolution is "scientific."

How did life originate?

The simples living cell needs several hundred proteins. Even if every atom in the universe were an experiment with all the correct amino acids present for every possible molecular vibration in the supposed evolutionary age of the universe, not even one average-sized functional protein would form, creationists state. So how did life with hundreds of proteins originate just by chemistry without any intelligent design?

"Nobody knows how a mixture of lifeless chemicals spontaneously organized themselves into the first living cell," evolutionist professor Paul Davies admits. Andrew Knoll, professor of biology at Harvard University, said, "we don't really know how life originated on this planet."

How did the DNA originate?

As we know by now, life is based on long information-rich molecules such as DNA and RNA containing instructions for making proteins, upon which all life depends. However, reading of the DNA and RNA to make proteins, and the replication of DNA or RNA to make new cells both depend on a large suite of proteins that are coded on the DNA and RNA. Both the DNA and RNA and the proteins need to be present at the same time for life to begin.

It is like the question, what came first, the chicken or the egg? What are the minimum requirements for a cell to live?

- A cell membrane separating the cell from the outside environment.
- A system to store the information and to make it available at any given time in the appropriate sequence.
- The ability to read the information and to control the timing of the production process. This involves over one hundred proteins and other complex factors.
- The ability to manufacture all of the cell's biochemical processes including the ATP synthase.
- The capacity to copy all the information and to pass it to the offspring for reproduction. A recent simulation of the process of a simple bacterium with "only" 525 genes would need over one hundred desktop computers and ten hours to accomplish.

Some of the evolutionists claim that the origin of life is "not a part of evolution." In reality, virtually every evolutionary biology textbook has a section on the origin of life in the chapters on evolution.

UC Berkeley has the origin of life included in its "Evolution 101" course. The particular section is titled "From Soup to Cells - the Origin of Life". Many high-profile defenders of the evolution theory, such as P.Z. Myers and Nick Matzke, agree that the origin of life is a basic part of evolution, as does Richard Dawkins.

"…we must concede that there are presently no detailed Darwinian accounts of the evolution of any biochemical or cellular system, only a variety of wishful speculations," says Franklin M. Harold, Emeritus Professor of biochemistry and molecular biology at the Colorado State University.

How could the mutation create a new species?
A mutation is nothing but an accidental copying mistake in the DNA code. They are known for their destructive effects, including over one thousand human birth defect diseases such as hemophilia.

Mutations are extremely rarely helpful. Knowing this, how can a piece of scrambling existing DNA information create a new biochemical pathway or hypercomplex nano-machines with countless components just to make evolution possible?

How could such errors, and mutations are errors, create about 3 billion letters of new DNA information to change a microbe into a new species?

Why is natural selection considered evolution?

What evolutionists call natural selection is the survival of a species more adaptable to the particular living conditions. As an example, plants or animals who can adapt to hot weather can survive in this climate, while others don't. Evolutionists call this the "survival of the fittest."

Why is this considered evolution despite no difference in the DNA? Huskies, best adapted to the bitter cold in Alaska, are still "dogs" and not a new species.

How did new biochemical pathways originate?

The biochemical pathways require multiple enzymes working together in sequence, up to thirty at the same time, often in a specifically programmed sequence. Did this happen just by accident without any instruction?

All living organisms appear to be purposely designed because there is no chance to be produced accidentally.

"We must concede that there are presently no detailed Darwinian accounts of the evolution of any biochemical or cellular system, only a variety of wishful speculations." (Evolution supporting biochemist Franklin Harold)

How did multi-cellular life originate?
The main question is how did cells adapt to individual survival "learning" to cooperate and specialize. This includes undergoing programmed cell death to create complex plants and animals?

Williams' "autopoiesis hierarchy" talks of the irreducible structure of the cell, and finds a universal example in the self-making of a cell. It describes five levels of organization in all living things that are needed for autopoiesis to occur:

- Perfectly-pure, single-molecule-specific biochemistry.
- Molecules with highly specific structures.
- Highly structured molecules that are functionally integrated.
- Comprehensively regulated information-driven metabolic processes.
- Inversely-causal meta-informational (information about information) strategies for individual and species survival.

The repair or maintenance strategies are integral for the survival of the adult multicellular individual because cellular selection operates with cell populations, including multicellular organisms, to select for the most reproductively aggressive cells. This needs to be controlled at the organismal level to maintain bodily integrity, Creation.com points out.

"Multicellular organisms could not emerge as functional entities before organism-level selection had led to the evolution of mechanisms to suppress cell-level selection." (Pepper et al)

How did sex originate?
The asexual reproduction is much more productive than the male and female duplication. How could non-intelligent physics and chemistry invent the

complementary systems needed to plan for future coordination of male and female sex organs?

Where are the missing transitional fossils?
No evolutional progress from a simple cell to complex living organisms is possible without transitional forms. Millions of them should exist. Incidentally, no transitional fossils can be found.

How can living creatures remain unchanged over millions of years?
The evolution theory claims a change from primitive life forms to complex ones. Assuming that the evolution theory is true, how can living creatures remain totally unchanged for such a long time?

How can blind chemistry create factors such as mind, intelligence, meaning, or morality?
For evolutionists, there is no higher power, spirituality, faith, or God. This begs the question, where did the human mind, intelligence, meaning, and morality come from?

Where are the scientific breakthroughs of evolution?
The evolution theory is widely praised as being "scientific." Strangely, no scientific breakthroughs, confirming the theory, appeared during the last one hundred years.

"In fact, over the last hundred years, almost all of biology has proceeded independent of evolution, except evolutionary biology itself. Molecular biology, biochemistry, physiology, have not taken evolution into account at all." (Dr. Marc Kirschner, chair of the

Department of Systems Biology, Harvard Medical School)

The main question is, why do schools and universities teach evolution so dogmatically, stealing time from experimental biology that benefits humankind?

Why is Creationism banned from our schools?
Not surprisingly, science confirms the Bible, which does not explain everything but nothing in the Bible has been proven wrong. (See my book "No more doubt – Science confirms the Bible"). At the same time, evolution fails to explain thousands of factors concerning our origin.

"Evolution is a religion. This was true of evolution in the beginning, and it is true of evolution still today." (Michael Ruse, evolutionist science philosopher)

"Darwinism is not a testable scientific theory, but a metaphysical research program." (Karl Popper, science philosopher)

If the unprovable evolution theory can be taught at our schools, why can't the most proven creationism be presented at the same time also?

Mystery #22:
The Charles Darwin dogma

At the bicentennial of Charles Darwin's birthday and 150 years since his revolutionary book "On the origin of species," it is the perfect occasion to summarize his still hugely influential work and the consequences we still suffer under.

Despite unforgivable flaws, Darwin's evolution theory is one of the most influential movements of the 19th, 20th, and 21st centuries. It greatly shaped our modern philosophy, theology, sociology, biology, physics, chemistry, astronomy, and politics. Not to mention, destroying the belief in God for millions of people worldwide. Darwin even motivated me to become an agnostic after my High School years. After studying the Genesis and starting to compare his teachings to natural sciences, I recovered fast.

While secularists in science, culture, politics, and media still praise Darwin for his contribution to humanity, the damage he has done to the world is immeasurable.

His theory might be interesting reading, especially if you are desperately trying to find arguments against God and his creation, but scientifically it is science fiction at best or junk science at worst. It is worth summarizing Darwin's worst blunders.

Life arising from some "warm little pond"
Darwin assumed life was created "spontaneously" from dead matter such as dirt and warm water. Not surprisingly, life turned out to be supremely more complex than we can ever imagine.

Over a generation ago, the highly anticipated "Miller-Urey experiment" was supposed to recreate the origins of life. They were running a mixture of gases through heat and electricity, produced a tar-like substance that formed some kind of amino acids. However, the experiment was rigged, since oxygen, which was excluded, would have ruined the results. As we know now, oxygen was present when life first appeared.

Still, there was no assembly of amino acids in order to reach the next level of the building blocks of life. Missing were the unimaginably complex proteins, which also must be precisely integrated into very sophisticated biosystems.

"A junkyard contains all the bits and pieces of a Boeing 747, dismembered and in disarray. A whirlwind happens to blow through the yard. What is the chance that after its passage a fully assembled 747, ready to fly, will be found standing there? So small as to be negligible, even if a tornado were to blow through enough junkyards to fill the whole Universe." (The Intelligent Universe, 1983) As the law of biogenesis concludes, life is an extremely complex biological process and can only arise from intelligent life.

The simple cell organisms
Charles Darwin considered the crude, primitive drop of matter called protoplasm an easy building block of life. Nowadays, we know that even "simple" bacteria contain complex molecular machines. Each bacterium appears more like a sophisticated automobile factory

with multiple robotic devices and a complex computerized control center.

What are the main characteristics of a molecular machine?

- Data processing, storage, and retrieval.
- The artificial languages and their decoding systems.
- The error detection, correction, and proofreading devices for quality control.
- Digital data-embedding technology.
- The transportation and distribution systems Automated parcel addressing (similar to zip codes and USPS labels).
- The assembly processes employing prefabrication and modular construction.
- The self-reproducing robotic manufacturing plants.

The theory of "pangenesis"

Darwin thought that a primitive cell just needs some very basic information to create life. In reality, inside the nucleus of each human cell are found thousands of carefully codified instructions, called genes, that have to be translated, transported, and reproduced.

All information, scientists have realized, is not made of physical matter. It has no mass, length, or width but it can be conveyed by matter. Neither has it been shown that information can somehow "evolve or be improved through mutations."

Biology teaches us that each human DNA molecule contains some three billion genetic letters. Incredibly, the error rate of the cell, after all the molecular editing machines fulfilling their functions, is only one copying mistake, called a point mutation, for every 10 billion letters.

Jonathan Sarfati, a physicist, and chemist, explains: "The amount of information that could be stored in a pinhead's volume of DNA is equivalent to a pile of paperback books 500 times as high as the distance from Earth to the moon, each with a different, yet specific content. Putting it another way, while we think that our new 40 gigabyte hard drives are advanced technology, a pinhead of DNA could hold 100 million times more information" (DNA: Marvelous Messages or Mostly Mess?)

The intermediate fossils
It is absolutely logical that any lifespan of billions of years must produce countless fossils for each of the time periods. The evidence should show a fine gradation between the different animal species and have millions of intermediate links Darwin hoped for. Darwin himself wrote: "The number of intermediate and transitional links, between all living and extinct species, must have been inconceivably great. But assuredly, if this theory [of evolution] be true, such have lived upon the earth." (The Origin of Species, 1958)

Darwin conceded: "The distinctiveness of specific forms, and they are not being blended together by innumerable transitional links, is a very obvious difficulty... Why then is not every geological formation and every stratum full of such intermediate links? Geology assuredly does not reveal any such finely-graduated organic chain; and this, perhaps, is the most obvious and serious objection to my theory."

Otto Schindewolf, perhaps the leading paleontologist of the 20th century, wrote that the fossils "directly contradict" Darwin. Steven Stanley, a paleontologist who teaches at Johns Hopkins, writes in The New Evolutionary Timetable that "the fossil record does not convincingly document a single transition from one species to another'" ("An Evening With Darwin in New York")

The limited variety of species
Charles Darwin got his idea about "natural selection" mostly from observing artificial selection. As an example, he noted the way pigeon breeders came up with a big variety of pigeons. Darwin was very impressed by that but ignored that all the pigeons were still all classified as pigeons.

No one honestly disputes than animals and plants may change over long periods of time. However, this is only "micro-evolution", a mostly small change within a species. This is not the actual evolution, Darwin's proposed change from one species to a completely different one.

Darwinian evolution, as is taught in our schools, is "macro-evolution", or changes beyond the limits of the species kind to create another distinct species. Darwin's evolution is based on the following suppositions:

- All living things descend from a common ancestor.
- The principal mechanisms for the changes are natural selection and mutation.
- These are processes are unguided, a "natural" process with no intelligence at work behind them.

The French zoologist Pierre Grasse, a late president of the French Academy of Sciences, clearly stated that these adaptations "within species" actually have nothing to do with evolution. They are mere "fluctuations around a stable genotype." This is considered a "minor ecological adjustment."

Darwin also hoped we would discover fossils of millions of extinct animals with gradual transitions between them. As it turns out, Darwin lacked the basic understanding of the laws of inheritance and strict genetic barriers between species.

The Cambrian explosion
The most remarkable aspect of the evolution theory discussion is the Cambrian explosion, a sudden appearance of a variety of complex life-forms emerging suddenly, without any predecessors, at the very same low level of the fossil records. Logically, this discovery obviously did not confirm Darwin's evolutionary model of simple-to-complex life.

According to Darwin's evolution theory, all living creatures descended from one predecessor. As an analogy compared to human inventions, this would be like the washing machine, refrigerator, toaster, air conditioner, or a car all "evolved" from the same mechanical machine, such as a bicycle, without any transitional steps.

Instead of a few primitive forms of life, as Darwin hoped for, fossil records display an "explosion" of life, a wide variety of very complex organisms.

"So when you encounter the Cambrian explosion, with its huge and sudden appearance of radically new body plans, you realize you need lots of new biological information. Some of it would be encoded for in DNA - although how that occurs is still an insurmountable problem for Darwinists. But on top of that, where does the new information come from that's not attributable to DNA? How does the hierarchical arrangement of cells, tissues, organs, and body plans develop? Darwinists don't have an answer. It's not even on their radar.'" (Lee Strobel, The Case for a Creator)

After several generations of scientists committed to confirm Darwin's theory, there is still no evolutionary mechanism that can satisfactorily explain the sudden appearance of so many completely different life-forms.

The theory of homology
Darwin's belief in ancestry theory lead him to develop his homology theory. Consequently, if all living creatures share the same ancestor, humans can't be any exception.

Particular similarities in specifying living creatures such as five human fingers and five parts of the bat's wing let him believe that this fact is related to common ancestry. This "evidence" is what he called homology.

The repeating features in nature such as two eyes, two ears, and four legs are a logical repetition of good, reliable design. It points rather to an intelligent designer than a common forerunner.

Humans evolving from apes
For over a hundred years, humans really believed that they progressed from apes, some type close to a chimpanzee.

Despite some basic optical similarities, up to 99 percent of human DNA drastically differs from the genes of apes.

"Well, the new study concludes that the total DNA variation between humans and chimpanzees is rather 6-7%. There are obvious similarities between chimpanzees and humans, but also high differences in body structure, brain, intellect, and behavior, etc." (Science editor Stephan Anitei in "How Much DNA Do We Share With Chimps?" Softpedia).

Our laws of genetics don't ever allow a chimp to become anything but a chimp, or a man becomes anything but a man. Not even the slightest evidence of Darwin's belief was ever found.

The tree of life
As intriguing as the imaginary picture of the tree of life in Darwin's book at the first glance might be, it is a pure phantasy and defies everything we know and

can prove about biology. The tree of life is full of leaves with virtually no connection to branches and the trunk.

Darwin portrays an imaginary transformation of a common ancestor, billions of years ago, into a different species we see today. Actually, his drawings were only based on slight variations within a species after many generations, not the development from one species into a different one, as the evolution theory requires.

"The most fundamental problem of evolution, the origin of species, remains unsolved. Despite centuries of artificial breeding and decades of laboratory experiments, no one has ever observed speciation (the evolution of a species into another species) through variation and selection. What Darwin claimed is true for all species has not been demonstrated for even one species." (Dr. Jonathan Wells, The Politically Incorrect Guide to Darwinism and Intelligent Design)

Rejecting God and creationism
Charles Darwin was deeply influenced by his father Robert and his grandfather Erasmus. Both of them were staunch atheists. At the same time, he was also profoundly shaped by the Zeitgeist of the 19th century, which resulted in important social, political, philosophical, and religious upheavals.

Just about eleven years after the publication of his revolutionary theory, he admitted: "I may be permitted to say, as some excuse, that I had two distinct objects in view; firstly, to show that species had not been separately created, and secondly, that natural selection had been the chief agent of change..."

Darwin's main reason to develop and publish his theory was a religious one. He fundamentally rejected God and his creation. Instead, he promoted the idea that the world of matter and energy, mainly through natural selection and variation, might well account for all life we see around us. This philosophy is later known as scientific materialism.

Not surprisingly, Darwin never took a deeper look into the Book of Genesis and the history of creation.

"Darwinian evolution... makes me think of a great battleship on the ocean of reality. Its sides are heavily armored with philosophical barriers to criticism, and its decks are stacked with big rhetorical guns to intimidate any would-be attackers..." (UC Professor Phillip Johnson)

"But the ship has sprung a metaphysical leak, due to the growing case for intelligent design, and the more perceptive of the ship's officers have begun to sense that all the ship's firepower cannot save it if the leak is not plugged. There will be heroic efforts to save the ship, of course... The spectacle will be fascinating, and the battle will go on for a long time. But in the end, reality will win." (Darwin on Trial, 1993)

"Consider the eye 'with all its inimitable contrivances,' as Darwin called them, which can admit different amounts of light, focus at different distances, and correct spherical and chromatic aberration. Consider the retina, consisting of 150 million correctly made and positioned specialized cells. These are the rods [to view black and white] and the cones [to view color]. Consider the nature of light-sensitive retinal [a complex chemical]. Combined with a protein (opsin), retinal becomes a chemical switch. Triggered by light, this switch can generate a nerve impulse . . Each switch-containing rod and cone is correctly wired to the brain so that the electrical storm (an estimated 1000 million impulses per second) is continuously monitored and translated, by a step which is a total mystery, into a mental picture." (Michael Pitman, Adam and Evolution)

Mystery #23:
The common sense

The theory the evolution is dominating our public life all over the world. The public pressure is so dramatic that most people don't dare to object to evolution. The effect on our schools, colleges, public discourse, and political, ethical, and moral reasoning is virtually catastrophic.

Remarkably, it is not just the almost criminal deception by the progressive movement we face. It is also the total lack of logic and common sense the evolution theory is spoon-feeding our society.

"It is necessary to clearly distinguish the facts of life (the living beings, their organs, and functions, their metabolism, their reproduction, etc.) and the interpretation of these same facts. The evolutionary hypothesis imposes upon the facts a general paradigm, and the latter pretends to explain everything, as it gives an account of the origin - and therefore the truth—of all living beings. If this were really the case, the theory would be predictive and the rules of logic would apply to evolutionary biology as well as they do in any other field of science." (Dominque Tassot, ScienceVsEvolution.org)

The Science vs Evolution website summarizes the insurmountable problem progressive scientists have who adhere to the dogma of the evolution theory.

"The most evident breach of the rules of logic is a contradiction: the fact of affirming at the same time something and its contrary. This is flagrantly observed in four common theses of evolutionist speech."

- The contradiction between cause and effect is observable all over the evolution theory disregarding logic and common sense. "Evolutionism affirms that these same causes, able to produce today small variations inside species, long ago produced effects of a different nature or on a very different scale, such as the transformation of reptiles into birds or of land mammals into whales. Here lies a first contradiction, sufficient by itself to show that evolutionism is not a sound scientific thesis." (Dominique Tassot) "The stone that falls today cannot have lifted itself yesterday." (James Hutton)

- The continuity and numerical discontinuity between beings result in classification proceeded by the division into two contrasting parts (dichotomy). Specifically, it is impossible to affirm continuity between living beings, while admitting at the same time the evident discontinuities concerning the number of their organs.

- Darwin's basic thesis claims a "gradual" evolution through small "modification." In reality, the fossil record does not show a single intermediary being.

- Extrapolation is a widely accepted practice in science, whereby an assertion is extended beyond the framework within which observation has been made. Evolution, in fact, tries to extrapolate predictions to the contrary. The actual observations of living beings testify to the permanency of species. All variations always remain within the specified limits of each species.

- The shift of meaning confuses every topic whenever evolutionists avoid the precise definitions. With this trick it is possible, almost without noticing it, to "modify" the extension or the comprehension of the terms in use.

- Using the definition of species instead of variety confuses the common taxonomic rules. Science can't claim a common ancestor if there is no genetic link to prove this.

- Wrong differentiation between macro- and micro-evolution causes all kinds of problems. The inescapable confusion between two concepts designated with the same word credits the idea of macroevolution with all the proofs of well-attested microevolution.

- The "natural selection," as described by Darwin in his book "On the origin of species," insidiously suggests that nature imitates the selection of the supreme variation. He disregards the fact that any kind of "natural selection" is an intelligent act that cannot happen by chance.

- Living organisms often adapt to changes in their living space. This happens through mutations and does not constitute any kind of "evolutionary adaption" producing new organs. They may just influence the pre-existing organs of the creature and is not evolution.

- The circular reasoning is one of the most ridiculous arguments found in the arsenal of the evolutionist's arsenal. "Circular reasoning is a logical fallacy in which the reasoner begins with what they are trying to end with. The components of a circular argument are often logically valid because if the premises are true, the conclusion must be true." (Wikipedia)

- The "survival of the fittest" is a simplistic assumption. It defines biological fitness by the capacity of survival. However, nothing new or scientifically convincing has been added to that postulate.

- Following the stratigraphic scale, geological layers are classified by attributing to each period or each level "index fossils." The evolutionary conclusion is that the more ancient fossils will be found in more ancient rock, and more ancient rock is supposed to show more ancient organisms, in accordance with the highly problematic "tree of life."

- The burden of proof is one of the most problematic fields in the evolution theory. For example, for paleontologists, the proof can be found in biology, for biologists, in geology; and for geologists, in paleontology.

- Very common in the evolution field is an open refusal to make decisions or distinctions. They are constrained by the iron rule of their paradigm, brush aside the application of this principle and condemn themselves to allow contradictory positions to co-exist within their thinking.

- Gradualism implies a continuity between the forms of life: morphological, physiological, and genetic continuity. The contrary of gradualism is discontinuity. The science of paleontology established the absence of gradualism between the different forms of life. Even though, this assertion is now considered as "definitely proven." The hunt for the existential "missing links" has mostly stopped. Don't' we need the missing links?

- Evolutionists substitute the actual "evolution" from one species to a new one with mutationism, variations within one particular species. "Mutations confirm the permanence of the species as such, with its internal variability. But evolutionists reject this affirmation." (Dominique Tassot)

- Despite any kind of scientific proof of their theory, evolutionists consider their evolution hypothesis "final." The response to objections is mostly approached with a change of vocabulary, which doesn't answer anything but confuse the matter even more.

- "Most errors men make do not result from the fact that they reason wrongly on the basis of true principles, but from the fact that they reason rightly on the basis of false principles or of inexact considerations," Fenelon states. False promises make everything even worse.

- The disappearance of particular mentioned species also is an unanswered question.

- According to Darwin, evolution is always described as a "progressive process" which improves living beings. In reality, the exact contrary is observed: mutations are either regressive or neutral, and the formation of subspecies or races is always a form of specialization, which impoverishes the gene pool and limits the intermixing of genes.

- Since Darwin was not able to explain how all living organisms descend from one ancestor, a primitive cell structure, to the most sophisticated living creature like humans, he turned to the factor of time. What can't be done in real-time, must be able to be accomplished in millions or billions of years. This is why Darwin badly needed the theory that our universe is billions of years old.
- Highly problematic are also his "genealogical trees." A tree is supposed to be composed of a trunk, branches, and leaves. In Darwin's world, virtually only leaves exist, representing fossils and species. The rest of the tree is nothing but a totally artificial construct. The suggested "common ancestors" can be found only in naturalists' imagination.

"It is necessary to clearly distinguish the facts of life (the living beings, their organs, and functions, their metabolism, their reproduction, etc.) and the interpretation of these same facts. Now, if the intellectual products of evolutionism show so many logical anomalies, it becomes legitimate to question not only such and such a premise or paralogism, but the legitimacy of the paradigm itself." (Dominique Tassot)

Mystery #24:
The Textbooks

Throughout Europe and North America, as well as in many other areas of the world, the evolution theory is taught as a biological fact. Considering the inability of Darwinism to prove even the most basic foundations of their hypothesis, this is not just factually wrong but it is a clear scientific fraud.

Why do we let this happen? Why do we allow creation-rejecting progressives to poison the minds of our children? We all have to accept some blame because we either don't understand the problem or because we are so careless and inconsiderate?

What can we do? One thing is sure, we can't let this to continue because what we believe really matters. Darwin suggested that life began when chemicals in a pond somehow mixed together to spontaneously create living matter, even though he admitted it could not be proven.

Today's evolutionists cannot prove how life was first formed either. However, we know the answer. It is our sacred obligation to make it public, to defend the creationism, and to reveal the scientific fraud spoon-fed to our children and the society at large.

Mostly, our children may go to church with us. They even may know you don't believe in evolution. But they're not likely to really understand why evolution is wrong unless you talk about it with them specifically. What must we do?

Educate yourself about the issues

Nobody expects you should become an expert in biology, astrophysics, or geology, but you must have some general understanding of the evolutionary theory, as opposed to the creation, to discuss this with your children.

You must know the basic terms like survival of the fittest, speciation, spontaneous generation, common descent, random mutation, natural selection, and more. You should definitely understand what these terms mean and how they fit into evolutionary theory. This will allow you to discuss the issues on a smart and intelligent level.

You should raise questions such as:

- Does God really exist?
- What is creation?
- What is evolution?
- Does it matter what you believe?

There are many very useful sources such as the Institute for Creation Research (icr.org) or answeringgenesis.org. Be careful though, you can find on the web many different versions and interpretations, among them the "old-earth creationists." This category is "progressive creationism," which suggests that the six "days" of creation in Genesis do not refer to literal 24-hour days, but rather epochs that could be millions or billions of years in duration. People who believe that want to combine evolution with creationism where is no common ground. They are just not compatible.

Address the evolutionary theory head-on
As soon as you have the necessary understanding of the matter, plan an appropriate discussion with your kids. Do not wait for questions children might have, they often don't ask at all. Very good is a regular family meeting environment in your house, but this also could be a good discussion topic at your dinner table.

You may start the talk by asking the children what they know and whether they have any questions. Be careful, don't do all the talking! Try to ask questions such as what do they think about the claim that all living creatures descended from a single primitive organism.

Ideal for such a discussion always is a family trip. You may want to go hiking with them searching for fossils. Every fossil ever found is a fully formed and functional species. You could tell them that professional paleontologists have not found transitional fossils either. This is an ideal setting to explain how fossils found to deny the evolutionary claim of billions of years old earth.

Of course, the kind and level of discussion must always correlate to the age of your children. With small children, you might consider going to the zoo, an excellent place to talk about the natural selection and whether humans are descendants of apes, as Darwin claimed.

Explain the biblical creation
It is not enough just to explain what is wrong with the evolution theory, it is even more important to teach

your children about biblical creation. Start at a very young age. Read to them Genesis 1 and 2 until this information is firmly fixed in their minds.

Don't stop at Genesis. As we know, the Bible contains many verses confirming the Genesis account of creation, in particular in the books of Psalms and Isaiah. Try to read these to your child too. Debate these verses. Remind them what God did on each of the days of creation, and what it means for us today.

Field trips are the ideal occasion to highlight the nature and their creation. Explain to them that the wonders of nature can't be created by mutation and natural selection. Trips to a zoo are always a good idea. Your hands-on lessons will make the creation story much more real to your children.

Plan how to confront evolution at a school
As much as your kids are convinced that evolution theory is wrong, they must face their teachers and their lessons presented as a 'scientific fact."

Talk to your kids about this problem and develop a strategy on how to address in the school. Assuming the teacher is open and accessible, try to submit the pertinent facts to that person and ask to present both sides to the class, evolution, and creationism.

In any case, it is advisable to talk to the local school board. Unfortunately, some of them are very Marxist-oriented, dogmatic, inflexible, arrogant, and condescending. Through their dogma, they want you to make aware that they are "morally superior" to you. It is not much you can do about that with the exception to vote them out during the next school board election.

When it comes to an actual test, kids usually have no choice but to answer questions the way they are taught to them. It's about the grades. Make them aware that they are displaying their knowledge presented to them, not their belief or opinion about the topic. Kids may preface their essay with a remark such as "in accordance with the subject as taught in the class, I am answering the questions as follows…"

Believing what really matters in life

Believing or rejecting the evolution hypothesis determines our entire worldview. If the universe and life on earth evolved by itself over billions of years, then the Genesis account of creation must be pure fiction and God isn't real.

That is exactly what many in our society want to believe in evolution. If there's no God, then there are also no absolutes about right and wrong and people are free to do whatever they want to do.

Our values and morals are derived from our Judeo-Christian civilization. Denying our Ten Commandments, what Marxists and other tyrants always do, means that there are no ethics, no morals, and no values we are supposed to adhere to. If there is no God, we are just a chunk of meat, bones, and blood.

This is, among many other factors, why progressives don't cherish human life. For them, abortion is just a technical procedure to get rid of the inconvenience of a child. For socialists and communists, unborn babies have no right to life, they are not "human" in an ethical, moral, or legal sense.

In reality, however, killing unborn babies is nothing but murder.

We must teach our kids to believe what really matters in life. No decent worldview is possible if you believe in the theory of evolution.

For any honest, authentic, and virtuous person, the decision is very easy. While the evolution can't prove a single one of their most basic claims, the Bible doesn't explain everything but it has never been proven historically, scientifically, or factually wrong.

The laws of nature

There are things in life nobody can deny, bend, manipulate, or change: the laws of nature. As illogical, naïve, or stupid it may sound, this is exactly what the evolution theory does throughout Darwin's highly controversial hypothesis.

Why? Applying the laws of nature, virtually no evolutional claim can survive the scrutiny of scientific proof. Their dogma is rooted in Charles Darwin's deeply-seeded motivation to "disprove" the creation theory and God. In short, proving his atheistic worldview was more important than acknowledging the laws of nature.

The evolution theory survived so only through denial and deception. In fact, Charles Darwin knew that he can't prove his theory. He even admitted this in his writing. Therefore, he escaped to other explanations "to make it work." He theorized, maybe it's possible to circumvent natural laws just by adding millions or billions of years to the process.

Who knows what might have happened in billions of years?

The evolution theory teaches us that all matter came into existence by itself. Later on, the living creatures also "made themselves." This sounds more like Greek mythology than a sound scientific explanation.

- The matter is self-originating from "nothing." According to the first law of thermodynamics, this is impossible.
- Living organisms were made from dead matter billions of years ago. No such a biochemical process is scientifically conceivable.
- Living creatures became "more complex" by themselves after billions of years. The second law of thermodynamics destroys this argument in seconds.

Not surprisingly, none of the principles of the laws of nature can ever be violated. Period.

"Even if one day we find our knowledge of the basic laws concerning inanimate nature to be complete, this would not mean that we had 'explained' all of inanimate nature. All we should have done is to show that all the complex phenomena of our experience are derived from some simple basic laws. But how to explain the laws themselves?" (R.E. Peieris, The Laws of Nature)

- The law of the originator says that the designer of any product must be more sophisticated than the product."
- The law of manufacture implies that the maker of a product must be more complex than the product.

"Facts are the air of science. Without them, a man of science can never rise. Without them, your theories are vain surmises. But while you are studying, observing, experimenting, do not remain content with the surface of things. Do not become a mere recorder of facts, but try to penetrate the mystery of their origin. Seek obstinately for the laws that govern them!" (Ivan Pavlov)

- The first law of thermodynamic tells us that the energy cannot be generated by itself or destroyed. The energy can be changed from one form into another, but the amount of energy remains the same.
- The second law of thermodynamics causes the qualitative degeneration of everything. It is also called "entropy."

The evolution denies the effects of the natural laws and even argue that their theory stands "above the natural law." How is this possible? Nobody can explain.

Summarizing the problem, there are basically six principles of evolution that are absolutely incompatible with the natural laws of nature.

- Evolution operates only upward, never downward.
- Evolution always operates irreversibly.
- Evolution always operates from smaller to bigger.
- Evolution always operates from primitive to more complex.
- Evolution operates always from less perfect to more perfect.
- Evolution is never repeatable.

"If your theory is found to be against the second law of thermodynamics, I can give you no hope; there is nothing for it [your theory] but to collapse in deepest humiliation.")Arthur S. Eddington, The Nature of the Physical World)

Mystery #26:
The effect on morals, values, and principles

The evolution theory has a devastating effect on our society. This reaches into the deepest aspects of our social life and our culture. For good or bad, Darwin's publications belong to the most influential books of the last two centuries.

"In the world of Darwin man has no special status other than his definition as a distinct species of animal. He is in the fullest sense a part of nature and not apart from it. He is akin, not figuratively but literally, to every living thing, be it an ameba, a tapeworm, a flea, a seaweed, an oak tree, or a monkey - even though the degrees of relationship are different and we may feel less empathy for forty-second cousins like the tapeworms than for, comparatively speaking, brothers like the monkeys." (George Gaylord Simpson, "The World into Which Darwin Led Us")

Marxist propaganda tool
The effect on our morals, values, ethics, and principles is extremely upsetting, disturbing, and harmful. It turns our Judeo-Christian worldview upside down. Not surprisingly, it instantly became the most effective propaganda instrument of the Marxist movements worldwide.

Evolutionists think if human beings are merely chemical accidents, why should we be so concerned about what they do? Therefore, why would an evolutionist be angry at anything one human being does to another if we are all nothing more than complex chemical reactions? There is no such thing as "good" or "bad," just different. For today's Marxists, the evolution theory is a perfect excuse to defy our morals and values.

Atheists claim we don't need God to be good people. We can create our own morals and values, apart from any religious belief. We decide what is good and wrong. But what are evolutionists' "values"? What is their moral code?

Most evolutionists, however, are aware that evolution does not provide any basis for morality. Summary: If evolution is true, then there can be no universal moral code that all people should adhere to. Evolution is law of the jungle.

"Darwinism consistently applied would measure goodness in terms of survival value. This is the law of the jungle where 'might is right' and the fittest survive. Whether cunning or cruelty, cowardice or deceit, whatever will enable the individual to survive is good and right for that individual or that society." (H. Enoch, Evolution or Creation)

Many people are not aware that morality and rules most humans adhere to have their basis in the Bible, specifically in the literal history of Genesis. Independently from biblical creation, morality has no justification.

"What does the unbeliever, person who rejects the biblical God, mean by 'good,' or by what standard does the unbeliever determine what counts as 'good' (so that 'evil' is accordingly defined or identified)? What are the presuppositions in terms of which the unbeliever makes any moral judgments whatsoever?" (Dr. Greg Bahnsen, Christian philosopher)

Some unbelievers still may classify their actions as good or evil, but they do not have any ultimate foundation for defining what is good and what evil really is.

Christianity and evolution are diametrically opposed. There is no compromise to be found. Evolution is a purely nihilistic philosophy, blatantly disregarding our highest morals and values such as human life.

Poisonous philosophy
Some people today do not appear to realize that the same poisonous philosophy, evolutionism, that justified killings under Hitler, also badly influenced the American abortion mindset.

According to documents released years ago, Joseph Mengele, the Auschwitz death-camp doctor known as the "Angel of Death" for his experiments on inmates, was allowed to practice medicine in Buenos Aires for several years in the 1950s. He worked as a "specialist in abortions," which were illegal.

At nobody's surprise, the devil spirit who extinguished life at Auschwitz would practice a similar grisly crusade on life in the womb.

Everything is a question of somebody's worldview. If you believe in evolution theory claiming humans are nothing special because they descend from the same ancestors as rats, there is no respect for human life.

For evolutionists, we are just like animals, created by chance millions of years ago.

There is no good or bad, just "different." With this kind of mindset, there is no surprise that a human being becomes what Margaret Sanger, founder of Planned Parenthood, became.

Birth control for Hitler fan Sanger was "nothing more or less than the facilitation of the process of weeding out the unfit." As an active eugenist, she defined the field as "the attempt to solve the problem from the biological and evolutionary point of view."

She wanted to change things "to the construction and evolution of humanity itself." Sanger advocated applying "a stem and rigid policy of sterilization and segregation to that grade of population whose progeny is already tainted, or whose inheritance is such that objectionable traits may be transmitted to offspring."

Revealing pro-choice tendencies, she went on to promote the notion of giving "certain dysgenic groups in our population their choice of segregation or sterilizations."

The evolutionary ideology helped to establish the philosophical framework of communism and atheism in Russia, China, Eastern Europe, Cambodia, North Korea, and others. Charles

Darwin's concept of the "survival of the fittest" has been repeatedly used to justify genocide against ethnic groups they considered "inferior."

Progressives around the world are in the process of changing their national laws to make eugenics legal. Eugenics is a set of beliefs to "improve" the genetic quality of a population by excluding people or ethnic groups considered "inferior" by promoting a "superior" race.

This corrosive, Hitler-style ideology also includes a physician-assisted killing of sick or otherwise heavily handicapped people whose life is deemed "not worth living." Only a secular, God-rejecting, evil mindset can promote eugenics. The excuse, again and again, is the evolution theory.

Our morals and values are not negotiable
For the wide majority of traditionally raised people, morals, values, ethics, and principles are very important. These societal rules determine how we should behave. They teach us all the fundamental issues of daily decision making, what is right, and what is wrong. A few examples:

- Is it right or wrong to clone a human being for medical research?

- Is it right or wrong to kill an unborn baby in the womb?
- Is it right or wrong to start a war for a good cause if still innocent people will be killed?

"Christian ethics ask what the whole Bible teaches us about which acts, attitudes, and personal character traits receive God's approval and which ones do not." (Wayne Gruden, Ph.D., Cambridge)

Throughout our history, ethics in Christianity are primarily centered on grace, mercy, and forgiveness. Every Christian is expected to have thoughts and deeds that are respectable and honorable and must abstain from doing any sins. The Ten Commandments are a perfect summary of Christian ethics.

What does Christianity teach us about morals and values, in addition to the Ten Commandments, evolutionists don't consider important?

- Worship only God
- Be kind to all people
- Be humble
- Be honest
- Live a moral life
- Be generous
- Say what you mean and mean what you say (Don't be a hypocrite)
- Don't be self-righteous
- Don't retaliate
- Be forgiving

Additionally, the following core values are typical for the Christian life:

- Grace
- Hope
- Faith
- Love
- Justice
- Joy
- Service
- Peace

In theory, these basic values sound very simple. In real life, however, the application of them often faces major hurdles because they demand goodness, self-discipline, and sacrifice many are not ready to commit to.

The building block of our worldview

Our "moral compass" is the irreplaceable part of our personality. It is the basic building block of our mindset and our worldview. You can't be a good person if you don't apply morals, values, and principles to your daily life.

- Most of us strive to live a purposeful life. However, if we are not deeply grounded in a moral code, we drift aimlessly through our earthly existence.
- Are you emphatic and care about other people who might be less fortunate? Only solid moral values help us to do care for others, to help others who are in need.

- We expect people to respect us. However, this is not a one-way street. Are we ready to respect others, too? Christian values demand that.
- We like to hear if other people call us a "good person." This, however, can't be taken for granted. Our attitude must reflect goodness and kindness at all times.
- Do you have a meaningful life? This very much depends on your priorities and your behavior. Only a resilient moral code can provide you that.
- Are you a source of "good things" in your family or the community? This always compels a strong moral code.
- According to research, good people usually are happier and healthier than others. Why? Physicians and psychologists view the main reason for that in your stable moral code and personal values.
- We all struggle with difficult decisions occasionally in our lives. Behavioral scientists, however, confirm that you can make better and wiser decisions if you depend on a genuine moral frame.
- Do you have problems trusting people sometimes? You can always easily trust others if you know that they are rooted in a solid moral code.

All this applies to our society at large, too. Shared moral values make us choose friendships and foster a firm sense of the community we belong to. Altogether, no happy, good life is imaginable without a foundation of morals, values, and principles. No mystery of the evolution theory can ever offer us that.

Our moral values are extremely important for our overall well-being. They provide the necessary structure for our lives.

- Honesty makes us respectable.
- Compassion makes us sympathetic to others.
- Courage gives us the bravery to overcome life's challenges.
- Modesty keeps us focused and humble.
- Forgiveness allows us to be emotionally stable because we don't hold onto anger and resentment.

These attributes will allow us to live a life in a way that reduces our stress levels. We can find peace and harmony in our lives.

Altogether, moral values allow us to live our life in a manner that we can be proud of. The bonds that we form with others are also more fulfilling if live according to honesty, compassion, courage, modesty and forgiveness.

References

- Accuracy in the Media
- American Center for Law and Justice
- American Center for Liberty
- Alliance Defending Freedom
- American Enterprise Institute
- American Greatness
- American Principles Project
- Americans United for Life
- Christian Coalition
- Campus Reform
- Cato Institute
- Claremont Institute
- Club for Growth
- Concerned Women for America
- Committee to unleash prosperity
- CommonSense.org
- Conservapedia.com
- Constitutional Accountability Center
- Creation.com
- David Horowitz Freedom Center
- Declaration of Digital Independence
- Federalist Society
- First Liberty Institute
- Focus on the Family
- Fox News Network
- Freedom Works
- Heartland Institute
- Heritage Foundation
- Human Events

- Human Coalition
- Judicial Watch
- Leadership Institute
- Liberal Propaganda Exposed
- Liberty Counsel
- Life Action
- Manhattan Institute
- Media Research Center
- National Marriage Project
- National Review
- National Right to Life
- NeverGiveUpYourDream.US
- Newsmax
- OAN – One America News
- OpenSecrets.org
- Operation Rescue
- Parental Rights Foundation
- Parents Television Council
- Prager University
- Psychology Today
- Reason Magazine
- SaveOne.org
- Spectator USA
- Susan B. Anthony List
- Tea Party Patriots
- The Federalist
- Townhall.com
- TrueTolerance.org
- Turning Point USA
- Victims of Communism

Books by Pierre A. Kandorfer:

- What is the purpose of my life? – Seeking a life worth living for
- A Love Letter to America – The Secret of the American Spirit
- What's up? American Cheers, Jeers, and Tears
- Good, Kind, and Happy – Open Secret To Our Life
- You Don't Know Who You Are
- End Game – When truth doesn't matter anymore
- Whom Can We Still Trust?
- Idiots from Hell – Defying Lunatics Among Us
- No More Doubt – Science Confirms the Bible
- Find Peace of Mind or Lose Your Mind
- Fight Back Manual – Last Bet Strategies for Survival of Western Civilization
- Liberals Hijacking America
- Clouds over Beverly Hills
- Etc.

**More information is available
@ NeverGiveUpYourDream.US
or PierreKandorfer.com**